JIM BOUTILLIER-FISHERIES AND OCEANS CANADA
YELLOWEYE ROCKFISH

A GUIDE TO THE ROCKFISHES, THORNYHEADS, AND SCORPIONFISHES OF THE NORTHEAST PACIFIC

JOHN L. BUTLER
MILTON S. LOVE
TOM E. LAIDIG

UNIVERSITY OF CALIFORNIA PRESS
BERKELEY LOS ANGELES LONDON

University of California Press, one of the most distinguished university presses in the United States, enriches lives around the world by advancing scholarship in the humanities, social sciences, and natural sciences. Its activities are supported by the UC Press Foundation and by philanthropic contributions from individuals and institutions. For more information, visit www.ucpress.edu.

University of California Press
Berkeley and Los Angeles, California

University of California Press, Ltd.
London, England

© 2012 by The Regents of the University of California

Publication of this book was assisted by National Oceanographic and Atmospheric Administration, National Marine Fisheries Service, Southwest Fisheries Science Center; University of California Sea Grant; and World Wildlife Fund.

Library of Congress Control Number: 2012941947

Manufactured in South Korea

21 20 19 18 17 16 15 14 13 12
10 9 8 7 6 5 4 3 2 1

Title page photo of black rockfish, *Sebastes melanops*: Clinton Bauder
Cover illustration: Amadeo Bachar www.abachar.com

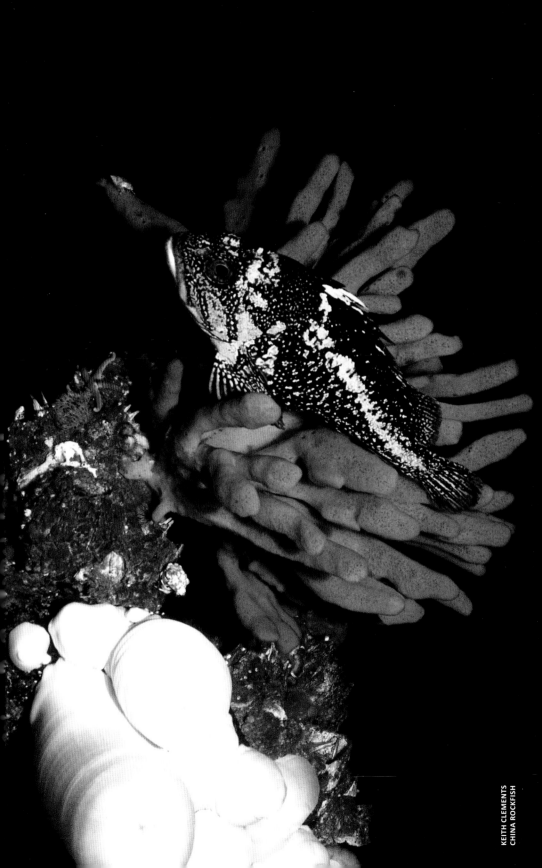

KEITH CLEMENTS
CHINA ROCKFISH

CONTENTS

Acknowledgments viii

About This Book ix

Rockfishes 1

Thornyheads 138

Scorpionfishes and … One Other Fish 146

Glossary 170

Appendices 171

References 181

Index 183

SWFSC ROV TEAM. SQUARESPOT ROCKFISH WITH PURPLE HYDROCORAL

ACKNOWLEDGMENTS

The creation of a book like this is, by definition, a team effort, an effort that extends well beyond the three authors.

We would like to thank Roberta Bloom, who tweaked the images and designed the book. Linda Snook, Andy Lamb, and Richard Rosenblatt pored over every species account and lent us their insights into the nuances of fish identification. Ken Franke, President of the Sportfishing Association of California, provided ship time and encouragement to write this book. Shauna Oh of University of California Sea Grant and William Fox of World Wildlife Fund helped us acquire funding for this book.

We are grateful for the efforts of Clinton Bauder, Scott Clark, Scott Johnson, Janna Nichols, Makoto Okamoto, Jay Orr, Ross Robertson, Chris Rooper, John Snow, and Carrie Worton, all of whom were patient and generous with their time and expertise.

We would also like to acknowledge the Southwest Fisheries Science Center's (SWFSC) remotely operated vehicle (ROV) team of John Butler, Scott Mau, David Murfin, Deanna Pinkard, and Kevin Steirhoff, for the many excellent images taken over seven years and more than 600 dives from Baja California to the Gulf of Alaska. The Southwest Fisheries Science Center, National Marine Fisheries Service, also supported the publication of this book.

Thanks also to the following for supplying many of the images: Schon Acheson, David Andrew, Takashi Asahida, Betty Bastai, Clinton Bauder, Arturo Ayala Bocos, Ed Bowlby, Cleo Brylinsky, California State University Monterey Bay (CSUMB), Marc Chamberlain, Tracy Clark, Keith Clements, Matt Cook, David and Dori Dirig, Steve Drogin, Scott Gietler, Richard Girouard, Greenpeace, Chris Grossman, Marcos Guimaraes, Stuart Halewood, Bernard Hanby, Dana Hanselman, Ruth Harris, Jon Heifetz, Peter Hergesheimer, John Hocevar, Mauricio Hoyos, John Hyde, Scott Johnson, Sal Jorgensen, Debbie Karimoto, Daiji Kitagawa, Donna Kline, Tom Laidig, Robert Lauth, Jon Lee, Kevin Lee, Robert Lee, Alberto Lindner, Steve Lonhart, Jim Lyle, Marine Applied Research and Exploration (MARE), Guy Martel, Neil McDaniel, Andy Murch, National Oceanographic and Atmospheric Administration Olympic Coast National Marine Sanctuary, Neptune Canada, Janna Nichols, Mary Nishimoto, Alexey Orlov, Jay Orr, Rick Ramsey, Rebecca Reuter, Hector Reyes, Dan Richards, Pauline Ridings, Ross Robertson, Axa Rocha-Olivares, Chris Rooper, Linda Snook, John Snow, Paul Spencer, Rick Starr, Duane Stevenson, Terry Strait, Brett Tischler, Stephanie Truhler, Bob Wohlers, John Yasaki, and Mary Yoklavich. James Boutillier provided us with access to all of the images taken during a Fisheries and Oceans Canada cruise to Bowie Seamount. Mary Yoklavich encouraged us to begin this project and has always been a source of steadfast support.

The Divebums website provided us with leads to many images and to the photographers that created them, and thanks to John Moore for maintaining that site.

And lastly, we thank Lisa Krigsman, who did not do much for this publication, but who whined loudly enough that we said she could be in the acknowledgments.

About This Book

The approximately 84 species of scorpaenid fishes (i.e., rockfishes, thornyheads, and scorpionfishes) inhabiting the northeastern Pacific help insure that the lives of many fish biologists will be exercises in decades-long humility.

While determining the species of this sprawling group of fishes on deck or in the laboratory can pose problems for anyone, these difficulties particularly afflict those of us who conduct underwater surveys. Even today, after a collective century of experience, one or another of the authors will pass around a crisp, sharp image of some rockfish peeking out of a crevice, or, even worse, just sitting right out in the open, and we will all agree that we don't know what species we are looking at. Oh, we will have our theories. And we will back it up with chatter about the number of pectoral fin rays, or the absence of some blotch or smudge on back or head, or the shape of a spine under the eye. But really, after all of this time working on these animals, when we view these fishes underwater, we are still sometimes mystified. And, if our emails are any guide, we are not alone in our frustration, as a week hardly passes without someone sending us a photograph of a rockfish with a message line reading "What is this?"

New management paradigms including marine protected areas and ecosystem-based management, as well as recovery plans for overfished stocks, require non-lethal monitoring methods to assess rockfish stocks. Increasingly these methods involve visual or optical techniques to determine species abundance and distribution. There are a great many scorpaenid species in the northeastern Pacific, and because many species can easily be confused, we have written this book to improve the quality of data derived from these surveys. To this end, we have included, within a species, many examples of variability in body morphology and color patterns associated with ambient light levels, habitat type, and maturation stage.

It is the continuing challenge posed by the identification of these fishes that inspired us to produce this book. Specifically, this is a guide designed to help differentiate species when fishes are observed underwater. In practice, this means that, except in those rare instances where we were at a semi-loss for what to say, we have not distinguished species using fin ray counts, eye diameters, and those other meristic and morphometric characters beloved by researchers who deal with dead fishes. Rather, here we emphasize those characteristics that will aid the underwater observer, characteristics that include body shape, color, and pattern, as well as behavior. Similarly, we have used images of dead fishes only sparingly and only when sufficiently diagnostic underwater images are unavailable. In addition, we focus our attention on those life stages that have settled out of the plankton. In practice, this means that this book is primarily a guide to differentiating both adult fishes and those juveniles that have recruited to benthic habitats. Physical identification of scorpaenid larvae is in its infancy and identification of pelagic juveniles would occupy a volume of its own.

Why is it so difficult to identify these fishes?

MANY SPECIES ARE VERY CLOSELY RELATED AND THUS LOOK QUITE SIMILAR

Many of the scorpaenid species that live off our coast evolved less than a million years ago, some perhaps in only the last few tens of thousands of years, and some are just getting around to it. Many recently evolved species physically resemble their nearest relatives and identifying these fishes underwater, or even when one is in hand, can be a daunting task. In some instances, only a handful of physical differences separate species. As an example, to the eye, gopher and black-and-yellow rockfishes can be distinguished only by differences in body coloration. At the extreme, occasional individuals of some likely newly evolved species pairs (rougheyes and blackspotteds are a good example) cannot be differentiated using physical characteristics.

Given how closely related many species are, one might think that interspecies hybrids would occur with some regularity, further complicating identification. However, with the

Perhaps a brown and quillback hybrid? It has coloration reminiscent of a brown, but a dorsal fin perhaps more like a quillback.

exception of three species in the Puget Sound region, this does not seem to be an issue. In that region, brown, copper, and quillback rockfishes mate with some frequency and the resulting hybrids can be a challenge.

MANY SPECIES CHANGE APPEARANCE AS THEY MATURE

The fishes discussed in this book change body shape, color, and color pattern with age. Clearly, all rockfish larvae differ from their older selves and, with only a few exceptions, only genetic techniques are able to separate them to species. In time, these larvae transform into pelagic juveniles. These are somewhat more readily identified, using a combination of body color and counts of fin rays, gill rakers, and lateral line scales and pores. When these juveniles take up a benthic existence (or at least one more closely associated with habitat, such as with drifting kelp mats), they develop more distinctive characteristics. However, even here identification issues abound, as the early benthic juvenile stages of closely related species often look very similar. Newly settled juveniles of the "KGB" complex (comprised of kelp, gopher, black-and-yellow, and copper rockfishes), for instance, settle to similar habitats and are essentially impossible to tell apart underwater. Another difficult group is the "black-spot" complex (comprised of black, blue, canary, olive, widow, and yellowtail rockfishes). In this instance, subtleties in such characters as the amount of black pigment on the dorsal fin, fin color, and body saddling do provide some assistance, although using these can be difficult when a scuba diver is viewing small and active fish in turbid waters. Fortunately, within a year or two of settling most northeastern Pacific scorpaenids (yelloweye rockfish are one exception) assume the general colors, patterns, and shapes of the adults.

Gopher? Copper? Black-and-yellow?

INDIVIDUALS MAY QUICKLY CHANGE COLOR AND PATTERN TO REFLECT ALTERATIONS IN BEHAVIOR AND HABITAT.

The ability to rapidly change color and pattern (often in just a few seconds) is most obvious in those species that routinely both swim in the water column and rest on the sea floor (e.g., bocaccio, chilipepper, halfbanded, shortbelly, squarespot, stripetail, and widow rockfishes). In these species, fish lying on, or adjacent to, substrate are heavily mottled, spotted, and blotched. That same individual swimming in the water column lacks most or all patterning and is often drab. We have also noted, albeit to a somewhat lesser extent, this same capability in a number of species that only live in close association with the sea floor.

All of these color and pattern changes occurred within a few seconds. (Top: chilipepper; Middle: chameleon rockfish; Lower: blackgill rockfish). All photos taken by SWFSC ROV Team.

In some instances, scorpaenids respond to the specific type of sea floor habitat they occupy, becoming more patterned if associated with a more complex substratum. As an example, on soft sea floors, the base color of California scorpionfish tends to be brownish or reddish, while fish living on reefs encrusted with coralline algae are often heavily marked or streaked in purple. Similarly, while the color of copper rockfish ranges from almost white to blackish, those living among coralline algae are often red. Lastly, although this had not yet been thoroughly studied, our observations suggest that there may also be geographic differences in the appearance of some scorpaenids. Rosethorn rockfish, for instance, seem to be darker and greener off Washington than off central California. On the other hand, the colors and patterns of adults of numerous species appear to be relatively stable. This is particularly true of many or most nearshore taxa (e.g., black, black-and-yellow, brown, China, grass, kelp, and olive rockfishes, and treefish). We also should note that color and pattern may change after death. Compared to living and unstressed individuals, freshly dead fish may be brighter or even have a different color and their pattern may diminish or disappear. Extreme examples are the chameleon rockfish and the various thornyheads. On the sea floor, they are often blotchy white and red (chameleons characteristically with red freckling on head and back). After capture, these fishes almost invariably turn completely bright crimson or pink.

Various scorpaenids may also develop abnormal coloration or patterns. By far the most common of these is *melanism*; black patches on the skin caused by abnormal growth of the black-pigment containing skin cells called *melanophores*. Similar abnormalities in other color-containing cells lead to red, orange, yellow, and white patches. Scorpaenids are also occasionally susceptible to color variations not related to skin cancers. For instance, golden bocaccio, vermilion, and widow rockfishes, among other species, have been reported and at least one albino broadfin thornyhead is known.

Blackgill rockfish.

Bocaccio.

Bank rockfish.

A LAST REALITY CHECK

When viewed underwater, most northeastern Pacific scorpaenids can be identified to species, given sufficient water clarity, enough time for careful observation, and a decent photograph or video frame grab. Ultimately, however, we should remember the guidance given by our associate, Lloyd Chen. When asked how he distinguished between several closely related and similar-appearing species, he replied "Sometimes you just have to go with the *gestalt*."

Lastly, ours is not the only reference to the identification of northeastern Pacific scorpaenids and we suggest consulting Eschmeyer et al. (1983), Orr et al. (2000), Love et al. (2002), Humann and DeLoach (2004, 2008), Robertson and Allen (2008), Lamb and Edgell (2010), and Love (2011) for additional assistance.

Juvenile yelloweye rockfish. MARC CHAMBERLAIN

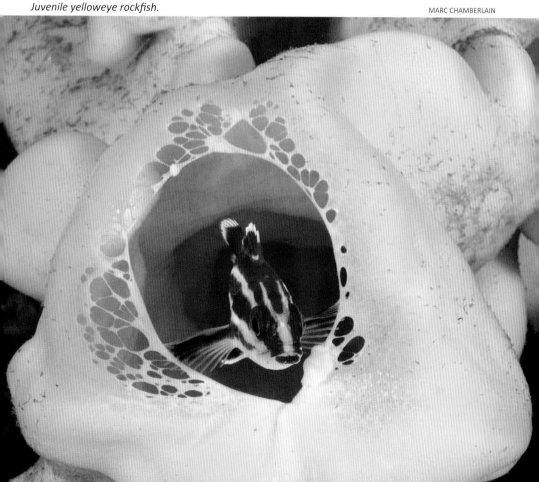

Rockfishes

BRETT TISCHLER
PELAGIC JUVENILE COWCOD

Sebastes jordani (Gilbert, 1896)
Shortbelly Rockfish

Maximum Length: 35 cm (13.7 in) TL. **Geographic Range**: Perhaps eastern Bering Sea (as far north as 54°58'N, 167°58'W) and north of Graham Island, British Columbia (54°22'N, 133°26'W) to southern Baja California (23°28'N, 110°43'W), most abundantly between Point Reyes, northern California and the Northern Channel Islands, southern California. **Depth Range**: 10–491 m (33–1,611 ft), primarily from 150–270 m (492–886 ft). **Habitat**: We almost always see shortbellies in large, sometimes very large, actively moving and polarized schools, although on rare occasions we do observe individual fish. These are mobile animals and they can range from many meters off the sea floor (as shallow as 10 m, 33 ft, below the surface) to lying right on the substrata. Juveniles recruit from the plankton to hard structures such as kelp beds, oil platforms, and rocky reefs. Adults do not seem to be wedded to any particular sea floor habitat. **Description**: Shortbelly rockfish are slim rockfish with deeply forked caudal fins. A key character is that, unique to this species, the anus is located midway between the anal and pelvic fins. In the water column, shortbellies are silvery or light brown, with relatively faint saddles and light specks along the sides. When near or on the bottom, they are more intensely colored, usually red or red-brown, with darker saddles and numerous white spots and blotches on the head, back, and sides. In many, or perhaps most, instances the tips of the dorsal spines are white.

SIMILAR SPECIES
Underwater, young juvenile bocaccio (smaller than about 10–12 cm, 4–5 in, long) swimming in the water column can be very difficult to separate from shortbellies, as the black spots and larger mouth of the larger juveniles may not be as obvious in smaller individuals. However, Scott Clark has observed that "shortbellies maintain tight, organized schools that move in unison much like anchovies," while schools of juvenile bocaccio are more amorphous. Chilipeppers have a similar body shape and are often mottled with red, but lack the white spots along the lateral line. Benthic chilipeppers have three, red "/" marks on the flanks, while shortbellies have a mottled appearance without the bars.

NOTE
The eastern Bering Sea records that we report are based on collections in the University of Washington Fish Collection (i.e., UW 134206). We have not verified the identities of these specimens.

Midwater appearance.

CSUMB MARE

Benthic appearance.

Sebastes goodei (Eigenmann & Eigenmann, 1890)
Chilipepper

Maximum Length: 59 cm (23.2 in) TL. **Geographic Range**: Pratt and Durgin seamounts, eastern Gulf of Alaska and northeast of Graham Island, British Columbia (54°37'N, 131°24'W) to off Bahia Magdalena, southern Baja California, typically from Cape Mendocino, northern California to at least Punta Colnett, northern Baja California. **Depth Range**: Near surface to 515 m (0–1,689 ft), mostly in 75–325 m (246–1,066 ft). **Habitat**: Juveniles recruit to nearshore kelp beds and other structure where they may form large schools. Adults, too, school in midwater over banks, although we have seen them as solitary individuals resting on the sea floor. **Description**: Chilipeppers are fairly elongate fish (although larger individuals develop a somewhat deeper body) with reduced head spines and a jaw that does not extend beyond the back of the eye. Their ability to rapidly change both color and pattern makes generalizations difficult. *Juveniles*: The first three anterior-most bars on the flanks are red and slant like so: "/" (when the fish is facing left, of course). *Older Juveniles and Adults*: When in motion or hovering in the midwater, older juveniles and adults are brown or pink-red on backs and pink on the sides, often with some brassy overtones.

Juvenile.

When at rest on the sea floor, chilipeppers are heavily mottled, blotched, and saddled with a mixture of red, salmon, pink, and white, with proportions varying among individuals. These individuals also have the three red "/" marks on the sides. We have also noticed that the spiny dorsal elements of both juveniles and adults tend to either be light or to be tipped in white.

SIMILAR SPECIES
The three, red "/" marks on the flanks of benthic chilipeppers appear to be distinctive to that species. However, when in the water column, shortbelly rockfish and smaller bocaccio can look similar. When a shortbelly's more anterior positioned anus is not visible, look for the slimmer body, smaller mouth, and more deeply forked caudal fin. Bocaccio have larger mouths (the jaw extending to the rear of, or beyond, the orbit) and, particularly in the young, dark spots.

CSUMB MARE

Note three red bars on the chilipepper.
That is a greenstriped rockfish above it.

Note white tipped spiny dorsal elements. Midwater color and pattern.

Sebastes paucispinis Ayres, 1854
Bocaccio

Maximum Length: 98.1 cm (38.6 in) TL. **Geographic Range**: Western Gulf of Alaska south of Shumagin Islands and Alaska Peninsula to Punta Blanca (29°05'N, 118°13'W), central Baja California and Isla Guadalupe, central Baja California. Larvae have been taken off Isla Cedros, Punta Eugenia, and as far south as about 26°N and 115°W in southern Baja California. These fish are most abundant from northern California to at least Bahia San Quintin, northern Baja California with some high-density pockets off British Columbia. **Depth Range**: Juveniles near surface and in inshore waters, adults in about 20–475 m (66–1,578 ft), most at depths of 95–225 m (312–738 ft). **Habitat**: Bocaccio are schooling, somewhat mobile, fish that live both on and near the sea floor and into the water column. Juveniles recruit to relatively shallow waters, amid such complex habitats as kelp beds, rocky reefs, and oil platforms. Adults tend to migrate into somewhat deeper waters and, although they primarily inhabit high relief areas, are also found over low relief and soft sea floor. **Description**: Bocaccio are elongate, laterally compressed fish with a large mouth (the jaw extends to the rear of the eye or beyond). Similar to both chilipeppers and shortbellies, bocaccio are capable of changing color and pattern with both age and habitat.

Young-of-the-year.

Juveniles: Newly recruited juveniles are red-brown to brown and may have a number of saddles and bars. Older juveniles have black, brown, or red-brown spots, some of which may remain into adulthood. *Adults*: Larger fish are salmon, brown, gray, reddish-brown, or red when in the water column, and develop conspicuous white markings on the back of the head, back, and sides when resting on the sea floor. While smaller individuals are fairly sleek, this species becomes quite deep bodied with age. **Note:** Declared "overfished" by the National Marine Fisheries Service and "threatened" by the Committee on the Status of Endangered Wildlife in Canada.

SIMILAR SPECIES

Underwater, and when in the water column, young-of-the-year bocaccio (less than perhaps 10–12 cm, 4–5 in, long) may resemble similar-sized shortbelly rockfish, as both species are metallic looking, can swim in very large schools, and the characteristic large jaw and dark spots of the bocaccio sometimes have not yet formed. In addition, the key character of the shortbelly, the anterior position of the anus, is usually not visible in rapidly moving individuals. Scott Clark notes that, even at a very young age, the two species differ in schooling behavior, as "shortbellies maintain tight, organized schools that move in unison much like anchovies while schools of bocaccio are looser." Linda Snook has observed that "the bodies of bocaccio are also shinier than those of shortbelly." Larger bocaccio are most likely to be confused with silvergray and Mexican rockfishes and chilipeppers. Silvergray rockfish are dark gray or green above, silvery on the sides, and white on the belly. They have a stronger symphyseal knob than do bocaccio. A white patch inside the mouth on the anterior of the inner lower jaw found in at least some silvergrays may also be diagnostic. Chilipeppers have a smaller mouth and lack spots. Chilipeppers on or near the sea floor have three "/" red bars on the sides that are absent in bocaccio. Mexican rockfish have smaller mouths and are darker on the back.

Juvenile.

Typical body pattern of fish lying on bottom.

Sebastes brevispinis (Bean, 1884)
Silvergray Rockfish

Maximum Length: 74.4 cm (29.3 in) FL. **Geographic Range**: Southeastern Bering Sea to Bahia de Sebastian Vizcaino, central Baja California, characteristically from Shumagin Island, Alaska Peninsula and southward to British Columbia. **Depth Range**: Surface to 580 m or more (1,902 ft) and mostly at 100–300 m (328–984 ft). **Habitat**: Juveniles sometimes associate with kelp beds and have also been found amid drifting algae. This is a schooling fish that favors complex habitats such as rock ridges, boulders, and vertical walls, but also occasionally cobble sea floors. **Description**: Silvergrays are relatively slim fish with reduced head spines, a relatively large mouth (including a fairly large symphyseal knob), and a projecting and massive lower jaw. Biologist Rick Stanley holds that the tip of the lower jaw of this species tends to be light in color. *Juveniles*: These are yellow-tan or greenish-brown, with a reddish wash to the ventral region. *Adults*: Adults are dark gray, green, or brown on back, with silvery or tan flanks. The lower part of the head, belly, and pectoral, anal, and pelvic fins may be washed with red. The body is not marked with spots or other markings.

Young-of-the-year.

SIMILAR SPECIES

At least at first glance, bocaccio and silvergrays are easily confused with one another. However, bocaccio are reddish, pink, or pink-brown, often have small dark spots, do not have as strong a symphyseal knob, and have a significantly larger mouth. Chilipeppers have small mouths, are more streamlined, tend to be pink or reddish, and often are blotchy or saddled with light and dark areas. Shortbelly rockfish are much more streamlined (nearly mackerel-shaped), have larger eyes, a smaller mouth, a strongly forked tail, and are often heavily saddled or otherwise darkly marked.

Juvenile.

SWFSC ROV TEAM

Silvergray on left and blackspotted on right.

JIM BOUTILLIER-FISHERIES AND OCEANS CANADA

Sebastes macdonaldi (Eigenmann & Beeson, 1893)
Mexican Rockfish

Maximum Length: 66 cm (26 in) TL. **Geographic Range**: Point Sur, central California to southern Baja California (23°24'N, 111°11'W) and Isla Guadalupe, central Baja California and the central part of the Gulf of California. Occasional in southern California, this species is abundant from central Baja California southward and in the Gulf of California. **Depth Range**: 76–350 m (249–1,148 ft). **Habitat**: Mexican rockfish are benthic fish that favor complex and hard sea floors. We have observed them on only rare occasions and it is unclear if this is a schooling species. **Description**: Mexican rockfish are a moderately deep-bodied species with reduced head spines and a jaw that does not extend behind the eye. *Juveniles*: Newly settled juveniles have dark saddles and bands that extend from the back downwards to the belly. The background color on the back is dusky and on sides is red (as is the lateral line). All fins except for the dorsal have crimson rays alternating with black. *Older Juveniles and Adults*: These are patchily dark on back and upper sides transitioning to reddish below, with a somewhat lighter belly. The lateral line is pink or pink-red, dark bars radiate posteriorly from under the eye and from between the eye and upper jaw, and white blotches may occur on the head and back.

SIMILAR SPECIES

Only bocaccio are likely to cause confusion with this species, as older juveniles and adults have a somewhat similar appearance. However, bocaccio have significantly larger mouths (with the jaw reaching to the rear of the orbit or beyond), are not nearly as dark on the back (they are pink, pink-brown, gray, or red) and often have small spots on the sides. Bank rockfish are more oval, do not have dark backs, usually have dark spots on the back and sides, and black in the fin membranes.

Sebastes rufus (Eigenmann & Eigenmann, 1890)
Bank Rockfish

Maximum Length: 55.2 cm (21.5 in) TL. **Geographic Range**: Queen Charlotte Sound (53°51'N, 130°45'W), British Columbia, to central Baja California (29°02'N, 118°13'W) and Isla Guadalupe, central Baja California. **Depth Range**: 31–512 m (102–1,680 ft), in southern California most often in 175–300 m (574–984 ft). **Habitat**: We primarily observe banks over high relief, in boulder fields and steep rocky slopes, and only occasionally over mixed rock-mud. While they often shelter, individually or in small groups, in crevices and caves, in some instances they form aggregations of dozens of individuals. **Description**: The bank rockfish is a deep-bodied, laterally compressed species with a protruding lower jaw. It exhibits a wide variety of colors and color patterns. *Juveniles*: Young-of-the-year are brassy or pink and have wide alternating dark and light vertical bars and a clear pinkish or white lateral line. *Older Juveniles and Adults*: These are various shades of brown, reddish, or almost white, often with dark and light saddles and blotches on the back and sides and usually (although not always) with a number of black spots on body and fins. An apparently reliable character is the whitish, pink, orange, or red stripe along the lateral line. There is a black band near the tips of the membranes of the spiny dorsal fin and the membranes of the soft dorsal, anal and pectoral fins are darkly pigmented. A dark "<" radiates from behind the eye.

SIMILAR SPECIES

At first glance, young-of-the-year speckled rockfish resemble similar-sized banks in shape and patterning. However, young speckleds are tan or yellow-tan with small black dots and lack solid pigment in the dorsal fin membrane. Larger speckleds have a distinctive symphyseal knob, sharper snout, lack the blotchiness of many banks, and are covered in fine black speckles. Semaphore rockfish have black marks on the dorsal fin, however they are not oval and lack the distinctive dark spots, saddles, and blotches. Bronzespotted rockfish have a strongly upturned jaw, larger spots, and a more squat body. Mexican rockfish have black backs, are more elongate, lack dark spotting on back and sides, and do not have the profuse black on the fin membranes.

Juvenile.

Sebastes ovalis (Ayres, 1862)
Speckled Rockfish

Maximum Length: 56 cm (22 in) TL. **Geographic Range**: Northern Washington (47°38'N, 121°56'W) to Arrecife Sacramento (29°40'N, 115°47'W), and Isla Guadalupe, central Baja California, typically from central California to at least Punta Colnett, northern Baja California. **Depth Range**: 30–366 m (98–1,200 ft), in southern California primarily in 90–140 m (295–459 ft). **Habitat**: A schooling species, speckleds are found both near the bottom and somewhat above it, over both high- and low-relief boulders and rock ridges. **Description**: Speckled rockfish are oval, laterally compressed, and sharp-snouted fish, with reduced head spines and a prominent symphyseal knob. *Juveniles*: Young-of-the-year are yellow-brown or buff, have somewhat darker vertical barring and, often, small dark spots. *Older Juveniles and Adults*: Larger fish are tan, brown, pink, or reddish-tan and have fine black speckles on the body and dorsal fin. When resting on the bottom, speckleds become quite blotchy or even may have alternating light and dark barring. The dark, fine spots, however, are always present.

Juvenile.

SIMILAR SPECIES
Juvenile speckleds may be confused with similar-sized banks, widows, and even squarespots. Young bank rockfish are pink, reddish, or brassy, have a clear lateral line, and, most importantly, their dorsal fins have distinct, dark black markings (all missing in speckled rockfish). Young widows are gray or brown and lack the symphyseal knob and the dark spots of this species. Juvenile squarespots are yellow-brown or beige, lack both the vertical barring and the dark spots, do not have a distinct symphyseal knob, and often have squarish pigmented areas along their backs. Larger speckled rockfish, with their combination of distinctive oval shape, very sharp snout, and dark speckles, are unlikely to be mistaken for any other northeastern Pacific species.

A blotchy one sitting on a rock.

A speckled rockfish on top and a bank rockfish on the bottom. Note only light speckling on the top fish.

Sebastes hopkinsi (Cramer, 1895)
Squarespot Rockfish

Maximum Length: 29 cm (11.5 in) TL. **Geographic Range**: Southern Oregon (36°38'N, 121°56'W) to northern Baja California (30°19'N, 116°06'W) and Isla Guadalupe, central Baja California, typically from central California to at least off Bahia San Quintin, northern Baja California. **Depth Range**: 18–305 m (60–1,000 ft), and primarily in 40–100 m (131–328 ft). **Habitat**: Both juveniles and adults of this schooling species live almost entirely over hard bottoms (ranging from boulder fields and rock ridges to cobbles, as well as around wrecks and oil platforms). During the night, and sometimes during daylight hours, squarespots shelter in crevices, while at other times they ascend well into the water column. **Description**: Squarespot rockfish are slim, oval-shaped fish with reduced head spines. *Juveniles*: Young-of-the-year are pale yellow-brown or beige and may not have the darker square spots characteristic of older fish. *Older Juveniles and Adults*: Larger fish range from yellow-brown or tan to dark brown and while they do have dark rectangular blotches on back and sides, these markings may be very faint in some individuals found in the water column. Fish that are near the bottom, either sheltering in crevices or in the open, are usually more heavily saddled and blotched, have a dark "<" mark behind each eye (this is often faint or even absent in midwater individuals), and also have a diagonal dark bar that extends over the back corner of the jaw (giving it a moustache appearance).

SIMILAR SPECIES
Particularly as juveniles, and particularly underwater, both widow and speckled rockfish may closely resemble this species. Juvenile widow rockfish are dark gray or brassy and never have rectangular blotches. Larger widows have sharper snouts, a "shiny", often

Juvenile. SCOTT GIETLER

brassy, appearance, and often a darker caudal fin. Juvenile speckleds are buff or yellow-brown with light barring and a dark-edged caudal fin and often have small dark spots. Larger speckleds are more oval than squarespots, have a sharper snout, and are covered in small, dark spots. Heavily saddled squarespots may also resemble the markings of sharpchin rockfish. Sharpchins are more elongate, have a more prominent lower jaw, and the blotch under the soft dorsal fin has a characteristic light spot. Dwarf-red rockfish are pink-tan, pink-gray, or dusky red, are slimmer, and have a whitish, pink, or reddish lateral line.

Blotchy pattern typical of a squarespot sitting on the bottom.

Note lack of obvious square spots on some of these individuals.

Sebastes entomelas (Jordan & Gilbert, 1880)
Widow Rockfish

Maximum Length: 62 cm (24.4 in) TL. **Geographic Range**: Albatross Bank, western Gulf of Alaska to Bahia de Todos Santos, northern Baja California, most commonly from at least southeastern Alaska to the Santa Barbara Channel, and on the offshore banks of southern California. **Depth Range**: Near surface (juveniles and adults) to 800 m (0–2,625 ft) and primarily in shallow waters (to the north) to 200 m (0–656 ft). Young-of-the-year rarely in tide pools. **Habitat**: This is a schooling species; it occupies both midwater and benthic habitats, usually over high relief. **Description**: Widow rockfish are relatively slim fish with reduced head spines. *Juveniles*: Young-of-the-year are gray or brown, often with a shiny and brassy finish and sometimes with tiny reddish spotting, and have darker saddles on back and sides. There is a weak dark spot on the rear of the spiny dorsal fin and fins range from clear to dark. Several stripes radiate from each eye. *Older Juveniles and Adults*: Somewhat older juveniles are brassy, gray, or brown and juveniles often sport light vertical bars and a dark caudal fin. Adults have similar colors and also have dark fin membranes and an overall dark caudal fin. While adults living in the water column usually do not have particularly distinct blotching or barring, individuals lying on the sea floor are often heavily marked with darker saddles. Andy Lamb reports that large adults often have a large pale blotch on each flank about midway between the head and tail.

Young-of-the-year. ROBERT LEE

SIMILAR SPECIES

Underwater, newly settled young-of-the-year widows resemble similar sized "blue rockfish" (composed of blue-sided and blue-blotched rockfishes). Young "blue rockfish" are somewhat deeper bodied, have wavy, red vermiculations (rather than the spotting young widows may have), and they do not have darker saddling. Scott Clark writes us "widows can be distinguished by their dark fin membranes. If ambient light is sufficient, the blue's orange mottling gives it an almost purple appearance while the widow typically appears gray. If lighting allows the diver to see these fish in silhouette, the anal fin of the blue forms a right angle while that of the widow is obtuse." Young-of-the-year squarespot rockfish, often found with young widows, are pale yellow-brown or beige, lack the dark fin membranes, and dark spiny dorsal fin spot, and may have faint versions of the rectangular side patches found in many adults of that species. When in mixed schools, the dark pigment in the fins, particularly on the rear edge of the caudal fin, is useful in distinguishing this species from similar-sized squarespots. Juvenile widows also school with young speckled rockfish. Juvenile speckleds are yellow-brown or buff, have darker vertical barring and, often, small dark spots. Larger juvenile and adult widows might be mistaken for similar-sized speckled or perhaps bank rockfishes. Speckled rockfish, however, are more oval, have a more pronounced snout and symphyseal knob, and are sprayed with black spots. Bank rockfish are similarly shaped to widows, albeit banks have larger symphyseal knobs. However, banks apparently always have a clear lateral line (generally red or pink), at least some dark spotting on the flanks, and, particularly, discrete and very dark marks on the upper spiny dorsal fin (widows may have some relatively faint dark marks).

Juvenile.

Juvenile.

Sebastes melanops (Girard, 1856)
Black Rockfish

Maximum Length: 69 cm (27.6 in) TL. **Geographic Range**: Southern Bering Sea and Amchitka Island, Alaska to northern Baja California, typically from at least Kodiak Island to central California. **Depth Range**: 0–366 m (0–1,200 ft), and mostly in 73 m (240 ft) or less. **Habitat**: This is a mobile, often schooling, species that can be found both on the bottom and well into the water column (occasionally leaping out of the water). Juveniles recruit from the plankton to shallow, nearshore (often intertidal) rocky areas, as well as kelp beds, algae, and eelgrass. Adults are most often found over high-relief rocky outcrops.

Description: The black rockfish is a fairly streamlined species with few head spines. *Juveniles*: Young-of-the-year are elongated, range in color from light brown to orange to red, sometimes have white spots along the dorsal surface, and have a dark spot on the rear part of the spiny dorsal fin. These young fish often have little in the way of patterning (the color tends to be fairly uniform), although a certain blotchiness on back and sides may be present. The caudal fin is typically clear or gray (matching the body color), but may be orange in orange individuals. *Older Juveniles and Adults*: These are bluish-black on back and sides with black speckles, a white belly, and white or gray blotches between the dorsal fin and the lateral line. The lateral line is often tracked by an uneven, thick white or gray stripe. Black rockfish have virtually no symphyseal knob.

SIMILAR SPECIES

Young-of-the-year yellowtail rockfish can easily be confused with similar sized black rockfish, as they are similarly shaped, have a prominent black spot on the spiny dorsal fin, and white markings on the back. There are, however, a few characters that seem to work in differentiating these

Young-of-the-year. CHRIS GROSSMAN

species. Small yellowtails are often heavily saddled, far more than are black rockfish. In addition, the black spot on yellowtails is larger. Lastly, many of the fins, particularly the caudal fin, are orange, regardless of the color of the fish. On a behavioral note, we have observed that, at least in the nearshore of central and northern California, young black rockfish are usually found from between the barely subtidal to mid kelp bed and usually very near rocks and algae (Scott Clark notes that young black rockfish often "hunker deep in crevices and display a body language that suggests nervousness or confusion.") Young yellowtails are more likely to inhabit mid kelp beds and deeper and stay more in the water column. Young-of-the-year olive rockfish also have the black blotch on the dorsal fin. However, that blotch is larger, the fins are orange or yellow, there are dark saddles on the back, speckling on the flanks, and larger white blotches on the back.

There are a number of relatively near-shore and shallow-water species that look similar to black rockfish. Dark rockfish have a similar body shape. However, that species is uniformly bluish-black to gray with only a gradual lightening in the belly region and has a distinct symphyseal knob (as compared to black speckling, pale blotches, and little or no symphyseal knob in black rockfish). The two species of "blue rockfish" are less elongate, have a bluish color, and a small mouth. They also lack the pale dorsal blotches of blacks. Dusky rockfish are greenish-brown, golden-yellow, or gray without the speckles and blotches of black rockfish and also have a strong symphyseal knob.

CSUMB MARE

MARC CHAMBERLAIN

JANNA NICHOLS

Sebastes ciliatus (Tilesius, 1813)
Dark Rockfish

Also in Asia

Maximum Length: 47 cm (18.5 in) FL. **Geographic Range**: Eastern Bering Sea and western Aleutian Islands to Johnstone Strait, British Columbia, commonly at least along the Aleutian Islands and Gulf of Alaska. **Depth Range**: 5–160 m (50–528 ft). **Habitat**: A schooling species, dark rockfish are found over high relief on rocky reefs and in kelp forests. **Description**: A complete description of this species and a comparison with the dusky rockfish are found in Orr and Blackburn (2004). Dark rockfish are semi-streamlined with reduced head spines. They may be more or less uniformly gray–black to gray or have black spots and splotches on backs and sides; the belly region is dark and does not contrast with the sides. They have a moderate-sized symphyseal knob.

SIMILAR SPECIES
Long considered to be the same species as the dark rockfish, dusky rockfish are typically lighter, varying from golden-yellow to greenish-brown to gray, with a distinctly white, pink, or red belly (dark rockfish bellies are dark) and a larger symphyseal knob. Black rockfish are easily confused with this species; they are black or bluish-black, with dark speckling on the back and sides, five or six light blotches on back, and a broad pale area along the lateral line. They do not have a symphyseal knob. "Blue-sided" and "blue-blotched" rockfishes (currently lumped under *Sebastes mystinus*) are blue or slate-blue and tend to have smaller symphyseal knobs.

We queried three experts on Alaskan juvenile rockfishes and two thought this was a juvenile dark rockfish.

SWFSC ROV TEAM

SWFSC ROV TEAM

SWFSC ROV TEAM

Sebastes variabilis Pallas, 1814
Dusky Rockfish

Also in Asia

Maximum Length: 59 cm (23.2 in) TL. **Geographic Range**: Hokkaido, Japan, and Kamchatka Peninsula. In the Bering Sea to about 60°N, Aleutian Islands and Prince William Sound to Johnstone Strait, British Columbia (less commonly southward of about Sitka, Alaska); one record from Oregon (44°24'N, 124°47'W). **Depth Range**: 6–675 m (20–2,228 ft), mostly from 100–300 m (328–984 ft). **Habitat**: This is a schooling species found primarily over high-relief sea floors.
Description: A complete description of this species and a comparison with the dark rockfish is found in Orr and Blackburn (2004). Dusky rockfish are fairly streamlined, with reduced head spines and a sharp symphyseal knob. Their backs and sides are usually golden-yellow to greenish-brown and, importantly, the belly area is white, pink, or red and contrasts with the color of the rest of the fish. Juveniles resemble adults, but have reddish-orange flecking on the sides.

We think that likely this is a juvenile dusky rockfish.

SIMILAR SPECIES
For many years dark rockfish were considered to be the same species as duskies. However, darks are more or less uniformly bluish-black to gray and the belly region is dark and does not contrast with the sides. They also have only a moderate-sized symphyseal knob. Black rockfish are black or bluish-black, with dark speckling on the back and sides, five or six light blotches on back, and a broad pale area along the lateral line. They do not have a symphyseal knob. The "blue rockfishes" have a generally blue or blue-blotched appearance with a small mouth.

SWFSC ROV TEAM

SWFSC ROV TEAM

SWFSC ROV TEAM

Sebastes mystinus (Jordan & Gilbert, 1881)
Blue Rockfish

Note: Both morphological and genetic research demonstrates that there are two species of "blue rockfish," here provisionally called "blue-sided" and "blue-blotched." Due to this confusion, the maximum lengths, as well as both geographic and depth ranges, of these species are not known.

Maximum Length: 53 cm (21 in) TL (both species combined). **Geographic Range**: Chatham Strait and Kruzof Island, southeastern Alaska to Santo Tomas (31°30'N), northern Baja California (both species combined). The northernmost range appears to be uncertain, at least partially due to these species' resemblance to the dark rockfish (*S. ciliatus*). **Depth Range**: 0–549 m (0–1,800 ft) (both species combined). **Habitat**: "Blue rockfish" are schooling reef fish. Particularly in nearshore areas, they can be found throughout the water column.
Description: Blue-blotched rockfish are blue with light gray or white blotches. The blue pigment is coalesced into darker blue blotches. They have a relatively small mouth and two prominent cheek bars, as well as bluish stripes across the front of their head. Blue-sided rockfish are similar to the blue-blotched but have relatively evenly pigmented flanks (compared to the dark blue blotches in the blue-blotched rockfish).

Young-of-the-year.

Blue-sideds have a more pointed snout, a more prominent lateral line (often light) and are more elongate. They have the relatively small mouth and two prominent cheek bars as with blue-blotched rockfish, as well as the bluish stripes across the front of their head. *Juveniles*: Both species have a weak spiny dorsal fin spot and are covered with red and blue vermiculations, and the fins are typically dark.

SIMILAR SPECIES
Particularly in low visibility or low light conditions, young-of-the-year widow rockfish closely resemble similar-sized "blue rockfish." Young widows are more streamlined, usually have at least some saddles on backs, and may be spotted (rather than vermiculated) with red. Adult black rockfish have a very distinctive mottled pattern, a light lateral stripe, and have a larger mouth. Dark rockfish are more or less uniformly gray–black to gray or have black spots and splotches on backs and sides; the belly region is dark and does not contrast with the sides. They also do not have the stripes across the head. Dusky rockfish are usually golden-yellow to greenish-brown on sides, with white, pink, or red sides and bellies, and a sharp symphyseal knob on the lower jaw.

Blue-blotched rockfish.

Blue-blotched rockfish.

Blue-sided rockfish.

Blue-sided rockfish.

Sebastes serranoides (Eigenmann & Eigenmann, 1890)
Olive Rockfish

Maximum Length: 61 cm (24 in) TL. **Geographic Range**: Southern Oregon to Islas San Benito, central Baja California typically from about the San Francisco area to southern California. **Depth Range**: Surface and intertidal to 172 m (564 ft) and mostly in 55 m (180 ft) and shallower. **Habitat**: Juveniles settle out of the plankton to nearshore structures, such as kelp beds, rock reefs, and oil platforms. Older juveniles and adults are found throughout the water column, either in small schools or singly, always close to some kind of structure. **Description**: Olive rockfish are relatively streamlined fish with reduced head spines. *Juveniles*: Young-of-the-year olives are elongated and have a series of dark saddles and white blotches on the back. There is brown specking on the sides and a large dark spot on the posterior part of the spiny dorsal fin. *Older Juveniles and Adults*: These are dark greenish-brown or brown on the back and a lighter greenish-brown, brown, or gray on sides, have greenish or light blotches on the back just below the dorsal fins, and greenish-yellow or drab-yellow fins.

SIMILAR SPECIES

Differentiating young-of-the-year yellowtail rockfish from olives of the same size can be a problem. In general, however, yellowtails lack the dark speckling on the flanks, have more square saddles, and tend to have orange (rather than yellow) fins. Young-of-the-year black rockfish are similarly shaped and also have that black dorsal fin spot. Small black rockfish do not have saddles on the back (they have a more uniform appearance), the black spot tends to be lighter and smaller, and the caudal fin is usually clear or gray. Underwater, both older juvenile and adult yellowtail rockfish look very similar to olive rockfish and we are not sanguine about always successfully separating the two species when in turbid waters. However, here are several characters that seem to work fairly well. Yellowtails are almost always more brightly and deeply pigmented. The back region is often dark brown, the light blotches often bright white, the fins bright yellow, the head with several yellow or green stripes, and the flanks have scales flecked in orange-brown or brown.

Young-of-the-year.

Older juvenile.

Three olives and a copper.

Sebastes flavidus (Ayres, 1862)
Yellowtail Rockfish

Maximum Length: 66 cm (26 in) TL. **Geographic Range**: Eastern Aleutian Islands south of Unalaska Island to Isla San Martin, northern Baja California. They are common as far southwards as central California and San Miguel and Santa Rosa islands and the outer offshore banks of southern California. **Depth Range**: Surface and to 549 m (0–1,801 ft), including intertidal zone, and typically in about 90–180 m (295–590 ft). **Habitat**: Juveniles recruit from the plankton to nearshore waters, settling among kelp and other algae, eelgrass, and around oil platforms. Both juveniles and adults are schooling fish, living in midwaters and near the bottom over high relief. On occasion, both juveniles and adults will rest on the bottom or seek shelter in crevices. **Description**: This is a typical slim midwater species with reduced head spines. *Juveniles*: Young-of-the-year are elongate and brown or orange-brown, with darker saddles and mottling on back and sides. Some or all of the fins, but particularly the caudal fin, are yellow or orange, and there are white spots on the back. Crucially, there is a prominent black blotch on the posterior part of the spiny dorsal fin. *Older Juveniles and Adults*: Older juveniles and adults are dark-brown or greenish-brown on back (above the lateral line) and light brown or tan flushed with yellow below that line and the scales are flecked with orange-brown or brown. The head often has yellow or green striping, particularly below the eyes. All fins are yellow or orange. There are a number of white or pale blotches just below the dorsal fins. The bright colors and blotches fade in death.

SIMILAR SPECIES

Young-of-the-year olive and black rockfishes can be easily confused with yellowtails. In particular, young olives and yellowtails are often quite difficult to differentiate, but here are several characters that may help: 1) Olives tend to have more yellow in the fins (they tend to be orange in yellowtails), 2) Olives have small, dark speckling on their flanks (yellowtails have blotching but not abundant speckling), and 3) Olives have

Young-of-the-year.

more rectangular saddling (they are more square in yellowtails). We have also noted that both juvenile and adult olives are more elongated than yellowtails, but the two species have to be observed in the same school to see this. Small black rockfish usually do not have dark saddles, the black spot is smaller, and the caudal fin is usually clear or gray. Similarly, older juvenile and adult yellowtails and olives may cause confusion. Olives are a generally drabber species, with dark greenish-brown or brown backs and lighter greenish-brown sides, greenish or light blotches on the back, and greenish-yellow fins. Importantly, they lack the abundant orange-brown or brown scale flecking and the yellow or green striping on the face of the yellowtail.

Sebastes alutus (Gilbert, 1890)
Pacific Ocean Perch

Also in Asia

Maximum Length: 75 cm (29.5 in) FL. **Geographic Range:** Southern Japan and Sea of Okhotsk to Bering Sea at Navarin Canyon, and Commander Islands and Aleutian Islands to Punta Blanca (29°08'N, 115°26'W), central Baja California. Larvae and juveniles may drift into the Chukchi Sea. They are most abundant in the northern Kurils and Kamchatka Peninsula, along the Aleutian Islands, in a few places in the southern Bering Sea, and southward to British Columbia. **Depth Range:** Near surface to 825 m (2,707 ft) and mostly in 100–400 m (328–1,312 ft). **Habitat:** Juveniles are found mostly on mixed sand and boulders in the shallower parts of the depth range. Larger fish tend to move into deeper waters. Adults often live over soft sea floors that are covered in sea pens, as well as over boulders and cobbles. Schools rise up off the bottom well into the midwater. **Description:** Pacific ocean perch (POP) are somewhat elongated fish with a long lower jaw tipped with a prominent and sharp symphyseal knob. *Juveniles:* The few juveniles that we have seen underwater were pink with extensive, somewhat darker blotches on back and sides (particularly a kind of linear blotch under the soft dorsal fin), a lighter pink lateral line, a dark blotch on the gill cover, and distinctive darker marks on the spiny dorsal fin. *Adults:* Most of the adults that we have observed underwater were light pink-brown or reddish, with narrow olive or brown patches on the back (the one by the soft dorsal fin is the most prominent), and the same dark patch on the rear of the gill cover (although this can be quite light) as found in juveniles. After death, they are red or pink. **Note:** Declared "overfished" by the National Marine Fisheries Service.

SIMILAR SPECIES

When viewed underwater, redstripe, sharpchin, northern, and yellowmouth rockfishes might be mistaken for POPs because of similarities in body morphologies and colors. Redstripes have a distinct gray, pink, or (after death) red stripe along the lateral line (POPs occasionally have a very diffuse and irregular clear area) and, at the most, have indistinct blotches on the back (those of POPs, particular under the soft dorsal fin, are dark and well-defined). Sharpchins have much more substantial saddling, at least several of these extending onto the flanks below the lateral line, and a more defined "<" radiating from each eye. In northern rockfish, the tips of the lower 8–9 unbranched pectoral rays are white, there is a dark "<" radiating back from the eye, and the caudal, soft dorsal, and anal fins are red. After death, yellowmouth rockfish also may closely resemble POPs. Yellowmouths have a weaker symphyseal knob, and yellow, black, and red blotches in the mouth.

Juvenile.

SWFSC ROV TEAM

Sebastes proriger (Jordan & Gilbert, 1880)
Redstripe Rockfish

Maximum Length: 61 cm (24 in) TL. **Geographic Range**: Pribilof Canyon, southeastern Bering Sea and Amchitka Island, Aleutian Islands to southern Baja California (26°46'N, 114°07'W), typically from southeastern Alaska to Oregon. **Depth Range**: 12–511 m (40–1,676 ft), mostly from 55–300 m (180–984 ft). **Habitat**: Redstripes school around boulders and rock ridges, cobbles and pebbles. Schools reportedly rise up off the bottom at night. **Description**: Redstripe rockfish are streamlined, with reduced spination, and a strong, dark symphyseal knob. *Juveniles*: The few juveniles we have seen were pink or greenish (on body and fins), sometimes with lighter blotches, had a clear lateral line, and lacked the strong symphyseal knob found on older fish. *Older Juveniles and Adults*: The clear lateral line of the early benthic juvenile is also found in older juveniles and adults, and is gray, pink, or (after death) red. Underwater, larger fish are red, pink, or tan, sometimes with vague darker mottling on back and sides, and there may be dark striations on the fins. The tips of the spiny dorsal fin membrane may be darkly edged, these spines may be tipped with white, and a "<" mark radiates from the eye.

SIMILAR SPECIES
Semaphore rockfish have dark marks on the upper parts of the spiny dorsal fin membrane. However, each of these marks is quite distinct (they are continuous in redstripes) and

Subadult.

semaphores are squat and lack the clear lateral line and the strong symphyseal knob. Harlequins are similarly shaped, have a lighter lateral line, and a "<" mark behind the eyes. However, they are darker and brighter, have quite dark fins that are often tipped with orange, lack a prominent symphyseal knob, and have a second anal fin spine that is longer than the third (it is shorter in redstripes). Sharpchins are also slim, have a strong symphyseal knob, and a "<" mark. They lack the clear lateral line. Pacific ocean perch have dark and well-defined blotches on the back and, occasionally, only a diffuse and clear area along the lateral line.

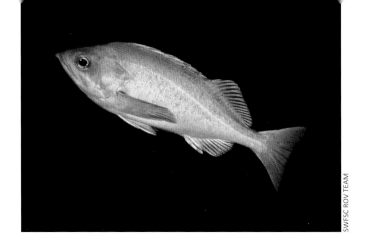
SWFSC ROV TEAM

LINDA SNOOK BETTY BASTAI

JIM BOUTILLIER-FISHERIES AND OCEANS CANADA

Sebastes zacentrus (Gilbert, 1890)
Sharpchin Rockfish

Maximum Length: 49 cm (19.3 in) TL. **Geographic Range**: Saint George Island, eastern Bering Sea and Attu Island, Aleutian Islands (52°55'N, 172°55'W) to San Diego, southern California and typically from about Kodiak Island to central California. **Depth Range**: 25–610 m (83–2,001 ft), but mostly from perhaps 200–300 m (656–984 ft). **Habitat**: While sharpchins occupy a wide range of habitats, they most often live near or on the sea floor on low- or high-relief hard bottoms or on mixed hard-and-soft sea floors. **Description**: Sharpchin rockfish are elongated fish, characterized by a large symphyseal knob, a dark "<" behind each eye (on occasion, this mark will be quite faint), a second anal spine that is longer than the third, and individual scales on back and sides are often visible.

Juveniles: Young-of-the-year are pink or tan with 4–5 brown saddles and without a prominent symphyseal knob. *Older Juveniles and Adults*: The basic body color of larger fish ranges from pink or tan to reddish with brown, orange, or red saddles on backs and often blotches on the sides. Note that very large sharpchins will, on occasion, have relatively small symphyseal knobs.

Young-of-the-year.

SIMILAR SPECIES

At least three rockfishes, harlequin and redstripe rockfishes, and Pacific ocean perch have somewhat similar body morphologies and/or color patterns. Unlike sharpchins, harlequins have red-, orange-, or clear-edged caudal and soft dorsal fins and a white, red, or orange lateral line along the posterior two-thirds of the body. They do not have a large symphyseal knob. Pacific ocean perch have dark marks on the back and a substantial symphyseal knob, but the marks do not extend down the sides and, unlike sharpchins, Pacific ocean perch have a dark blotch on the gill cover. Redstripe rockfish also have a large symphyseal knob, but they have a clear lateral line. Neither species has as much of a scaly appearance as does the sharpchin. Stripetail rockfish lack a "<" shaped mark behind the eye and onto the gill cover, have a smaller symphyseal knob, a less elongated body, and have greenish caudal fin striping.

Sebastes polyspinis (Taranetz & Moiseev, 1933)
Northern Rockfish

Also in Asia

Maximum Length: 48 cm (18.9 in) TL. **Geographic Range**: Eastern Kamchatka Peninsula and off North America, as far northward as Parvenets Canyon, Bering Sea (59°30'N, 178°W), to Graham Island, (42°N, 132°W), British Columbia. They are common all along the Aleutian Islands and Alaska mainland, perhaps to about the Yakutat area. **Depth Range**: 10–740 m (33–2,442 ft), mostly in 75–200 m (246–656 ft). **Habitat**: Northern rockfish are a schooling species. Young fish tend to live in relatively shallow waters and then move deeper as they mature. Most individuals live over high-relief bottoms and during the day they may school from near the sea floor to as much as 40 m (131 ft) above the bottom. At night, the fish tend to move downwards and shelter on the bottom. **Description**: Northern rockfish are elongate, fairly streamlined fish. After capture, or when a sharp photograph is available, they are easily distinguished from virtually all other rockfishes in the eastern North Pacific by having 14 dorsal fin spines. (Gray rockfish also have 14 dorsal spines, but that species is much different in body morphology and coloration). *Juveniles*: Juveniles are pale gray above, reddish below the lateral line, and pale ventrally. The blotchy red-brown spots of younger fish gradually coalesce in older fish. *Older Juveniles and Adults*: In larger fish, the tips of the lower 8–9 unbranched pectoral rays are white; we have found that, underwater, northern rockfish can be distinguished by this white mark at some distance. A darker "<" radiates back from the eye onto the gill cover. The caudal, soft dorsal, and anal fins are red; the spinous dorsal is pale with darker pigment at the tips of the spines.

Subadult.

SIMILAR SPECIES
No other northeastern Pacific species has the combination of red caudal, soft dorsal, and anal fins, white-tipped lower pectoral rays, and 14 dorsal fin spines. Pacific ocean perch lack the white-tipped pectoral rays.

ALBERTO LINDNER

SWFSC ROV TEAM

Sebastes melanosema Lea & Fitch, 1979
Semaphore Rockfish

Maximum Length: 39 cm (15.5 in) TL. **Geographic Range**: Questionable record off Oregon; Santa Barbara Channel, southern California to Punta San Pablo (27°13'N, 114°30'W), central Baja California. **Depth Range**: 97–490 m (318–1,607 ft). **Habitat**: This is a benthic species. While we have not seen many, they have always been solitary individuals, living on rocks and on soft sediment near rocks. **Description**: Semaphore rockfish are short, squat fish with distinctive dark markings on the upper parts of the spiny dorsal fin membranes and bifurcated lachrymal spines. They range in color from bright red with white blotches to pinkish-white with slightly darker bars.

SIMILAR SPECIES

Bank rockfish have dark pigment on the membranes of the spiny dorsal fins, but this species is more ovate and compressed, with a red lateral line and with black pigment often occurring in most or all fins. Redstripe rockfish may have black edging on their dorsal spine membranes. However, they are slimmer, are usually pink or brown with other markings, and have a clear lateral line. Splitnose rockfish juveniles also may have dark pigment in the dorsal fin, but have prominent dentigerous knobs and single lachrymal spines, larger eyes, and the body shape is more ovoid.

We think this juvenile is quite possibly a semaphore.

Note dark marking.

Sebastes variegatus Quast, 1971
Harlequin Rockfish

Maximum Length: 38 cm (15 in) TL. **Geographic Range**: Southeastern Bering Sea and Aleutian Islands at Bowers Bank to 95 km (59 mi) southwest of Newport, Oregon (44°32'N, 124°39'W**)**, commonly as far south as at least Dixon Entrance (between Alaska and Canada). **Depth Range**: 6–558 m (20–1,831 ft), mostly between 50–300 m (164–984 ft). **Habitat**: Harlequins are benthic species that most often occupy high-relief habitats. **Description**: These are streamlined fish with reduced head spines. A key diagnostic character is the lateral line, which is most differentially pigmented (white, pink, or orange) along its posterior two-thirds. Otherwise this is a red, orange, or pink-tan fish, with dark fins (the edges of the caudal and soft dorsal fins are orange, red, or clear), and often with heavy black or gray saddling or mottling on back and sides. The relatively few juveniles we have seen had backs that were darker and more heavily blotched with white than many larger fish. There is a dark "<" behind each eye.

SIMILAR SPECIES

The combination of a red, orange, or clear fringe on the caudal and soft dorsal fins and a distinctively pigmented lateral line are unique to this species. In addition, while the sharpchin rockfish has a similar body morphology and a "<" behind the eye, its symphyseal knob is more pronounced, the lower jaw is longer, it lacks most of the dark pigment in the fins, and the white, orange, or red lateral line. Redstripe rockfish have a clear lateral line from head to caudal fin and, again, lack the fin pigmentation of the harlequin.

Juvenile.

Another juvenile.

Two prowfish (Zaprora silenus) horn in on the picture.

SWFSC ROV TEAM

SWFSC ROV TEAM

SWFSC ROV TEAM

Sebastes saxicola (Gilbert, 1890)
Stripetail Rockfish

Maximum Length: 41 cm (16.1 in) TL. **Geographic Range**: Yakutat Bay, eastern Gulf of Alaska to Punta Rompiente (27°41'N, 115°01'W), southern Baja California commonly from British Columbia to at least southern California. **Depth Range**: 25–547 m (82–1,795 ft), off southern California, mostly in 180–270 m (590–886 ft). **Habitat**: Juveniles are found on sand, low-relief rocks, around sea pens, and the like. Juveniles that recruit to the edge of kelp beds and other shallow habitats tend to migrate into slightly deeper waters within a few months. Stripetails are usually solitary and tend to lie on the bottom or hover just above it; older juveniles and adults live over soft and mixed soft and low-rock habitats. **Description**: This is a fairly elongated species, characterized by a relatively large eye, greenish streaks along the membranes of the caudal fin, and a white patch at the base of the tail (the base of the caudal rays of smaller individuals are dark, producing a bar following that white patch). *Juveniles*: Young-of-the-year have a distinctive saddling, with one saddle stretching from the spiny dorsal fin to the anus and a saddle shaped like an inverted triangle under the soft dorsal fin. *Older Juveniles and Adults*: These range in color from red (particularly larger individuals) or red-brown to almost white, almost always with at least some darker saddles on head, back, and sides.

Young-of-the-year.

SIMILAR SPECIES
Halfbanded rockfish, particularly individuals that are heavily saddled and otherwise marked, can easily be confused with stripetails by the unwary. However, halfbandeds always have a unique and dark diamond-shaped mark just under the rear part of the spiny dorsal fin that extends to below the lateral line. Sharpchins have a "<"-shaped mark behind the eye and onto the gill cover, have a large symphyseal knob, and lack distinct, greenish caudal fin striping (they have some faint dusky coloration). The saddling pattern of stripetails can be confused for the upside down "U" pattern in splitnose (smaller individuals), especially in turbid or low light conditions. The striping in the caudal fin and the darker coloration of the saddles in stripetails can separate these species.

Juvenile.

SWFSC ROV TEAM

SWFSC ROV TEAM

SWFSC ROV TEAM

SWFSC ROV TEAM

Sebastes semicinctus (Gilbert, 1897)
Halfbanded Rockfish

Maximum Length: 25 cm (10 in) TL. **Geographic Range**: Strait of Juan de Fuca, Washington to Bahia Asuncion (27°01'N, 114°16'W), southern Baja California, and abundant from at least central to southern California. **Depth Range**: 15–402 m (50–1,320 ft), in southern California, they are mostly in 60–135 m (197–443 ft). **Habitat**: Juveniles recruit to low-relief rock and sand, and to the shell mounds surrounding oil platforms. Adults typically are found over mixed soft and hard low-relief bottom (like cobble beds), although they can also be found over rocks and boulders and over mud sea floors. We see them both as individuals lying on rocks or in little divots on mud, and singly or in schools some meters into the water column.

Description: Halfbanded rockfish are slim rockfish characterized by a dark and diamond-shaped mark that sits from just below the posterior part of the spiny dorsal fin to below the lateral line. Behind this dark feature (towards the tail) are two other bands that may or may not be dark, as this species is capable of changing pattern, and to a certain extent color, although the diamond

Young-of-the-year.

mark is always present. *Juveniles*: Juvenile halfbandeds are often silvery and, except for these three bands, only lightly marked. *Adults*: Larger individuals swimming in the water column (and dead individuals) tend to be orangey, reddish, or brown with or without some additional saddles and speckling. Fish of all sizes sitting on the sea floor tend to have considerable red on their flanks and are heavily saddled and blotched. Brownish or greenish streaks and dots line the tail of this species.

SIMILAR SPECIES

Stripetail rockfish look quite similar, in both shape and color, to this species and care has to be taken in order to avoid confusion. However, while stripetails may have a number of dark saddles, they lack that very dark, mid-body, diamond-shaped bar.

Two halfbandeds bracket a stripetail rockfish.

Sebastes diploproa (Gilbert, 1890)
Splitnose Rockfish

Maximum Length: 45.7 cm (18 in) TL. **Geographic Range**: Sanak Islands, western Gulf of Alaska to Isla Cedros, central Baja California, typically from British Columbia to at least southern California. **Depth Range**: (adults) 45–924 m (148–3,031 ft), typically from 200–420 m (656–1,378 ft). **Habitat**: Juveniles recruit to drift algae and other material and then settle out on mud and on low, but hard, relief. Older juveniles and adults live primarily on mud-rock and mud bottoms, however, splitnose have also been observed hanging on the sides of steep pinnacles. Although we observe most of these fish lying right on the sea floor (usually evenly spaced from one another), we have also, rarely, seen individuals and aggregations suspended well off the bottom. **Description**: Splitnose rockfish are deep–bodied rockfish with large eyes and an upper jaw that is notched or split with a prominent dentigerous knob on either side of the snout (these are most apparent in larger fish). The first (anterior-most) lachrymal spine is strong, single (undivided), and points forward and down. *Juveniles*: Pelagic juveniles dwelling under drifting material are light, tan, or brassy (often matching the surrounding kelp mat), sometimes with a yellow wash, and may have broad, slightly darker, vertical bars. Their large eyes distinguish them from other pelagic juveniles. *Settled Juveniles and Adults*: Newly settled juveniles are tan with darker bars; these darker bars turn red with time. The colors of older and settled individuals range from red and pink to tan, usually

The "split nose" is easy to see here.

(but not always) with at least some white blotching and mottling, with smaller fish often the most pale. In addition, individuals often have an upside-down "U" mark on the side (below the posterior part of the spiny dorsal) formed by two arches of color that originate at or near the ventral area. We have noted that the spiny dorsal fin of many individuals has a bit of dark coloration on the interspine membrane, near each spine tip. After capture, splitnose are almost always a uniform red or pink.

SIMILAR SPECIES

Underwater, both aurora and chameleon rockfishes might be mistaken for this species. An aurora rockfish lacks the strong dentigerous knobs on the upper jaw, has a single pointed lachrymal spine that aims backwards, and eyes that do not rise above the level of the cranium. Auroras typically are lighter than splitnose and instead of having an inverted "U" pattern on the upper flank, they have

a white blotch under the seventh through tenth spiny dorsal element. Chameleon rockfish also have large dentigerous knobs. However, the anterior-most lachrymal spine has multiple (rather than a single) points. Chameleons are often red or pink with larger pale areas that contain at least a few (and sometimes many) small reddish spots. They also lack the inverted "U" on their sides. Blackgill rockfish have dark pigment on the gill covers and do not have prominent dentigerous knobs. Small semaphore rockfish may be confused with splitnose. However, they lack dentigerous knobs, have smaller eyes, and a slimmer body shape.

Note upside-down "U" mark.

Note the black coloration near each spine tip.

Sebastes aurora (Gilbert, 1890)
Aurora Rockfish

Maximum Length: 41 cm (16.1 in) TL. **Geographic Range:** Southeastern Alaska (55°56'N, 135°26') to Isla Cedros, central Baja California (28°02'N). Larvae have been taken further southward off Banco Thetis (24°40'N, 112°18'W), southern Baja California. Typically from northern Oregon to at least southern California. **Depth Range:** 81–893 m (990–2,930 ft), most are found between 200–500 m (656–1,640 ft). **Habitat:** This is a benthic species, most often found over mixed hard and soft sea floors, but also occasionally over either sediment or low-relief rocks. **Description:** Aurora rockfish are relatively thick bodied (particularly across the head), spiny fish that have rounded snouts with no or poorly developed dentigerous knobs on the upper jaw; each lachrymal spine has one point and points rearward. *Juveniles:* Many or perhaps all juveniles have a dusky blotch on the gill cover and tend to have a number of amorphous red splotches on the head. *Adults:* Both juveniles and adults are red over much of the body, often with white or pale areas covering some or most of the top of the head, sides, and back. The lateral line can be white, pink, or pale red and the eye is large and sited below the top of the cranium. Importantly, there is a white patch at the base of the spiny dorsal fin below about the seventh through tenth spines. We expect this species is capable of changing its patterns. In death, auroras tend to be pink or reddish without distinct blotching.

SIMILAR SPECIES
When viewed underwater, we find that the adults of both splitnose and chameleon rockfishes may closely resemble this species. If visible, an easy way to differentiate splitnose from auroras are a splitnose's pronounced dentigerous knobs on the upper jaw and the first lachrymal spine that points forward and down. When these characters are not visible, note that splitnoses do not have a white blotch below the seventh through tenth hard dorsal spines (although this species often has other white blotches on the back) and the tops of the eyes rise *above* the cranium. Chameleon rockfish have a first lachrymal spine with multiple points and eyes that rise *above* the cranium. In addition, chameleons also often have distinct small, "measly" red spots on the pale areas (if these exist) of the head and back and on the fins.

SWFSC ROV TEAM

Juvenile.

Note white blotch.

Sebastes phillipsi (Fitch, 1964)
Chameleon Rockfish

Maximum Length: 52 cm (20.3 in) TL. **Geographic Range**: Point Saint George, northern California (41°34'N) to Nine Mile Bank (32°39'N, 117°28'W), southern California. **Depth Range**: 174–390 m (570–1,279 ft). **Habitat**: Chameleons are benthic and solitary fish that favor high-relief habitat. **Description**: Chameleon rockfish are deep-bodied, with particularly thick and spiny heads, two dentigerous knobs on the upper jaw (less apparent in small individuals), and eyes that bulge above the cranium. Along with the dentigerous knobs, another excellent diagnostic character (although one visible only in the sharpest photographs and video) is that the first lachrymal spine has multiple points. Underwater, these fish are usually pink or red, frequently with pale areas over the head, back, sides, and dorsal and caudal fins. These lighter areas, particularly those on the head, back, and fins, often have small red spots, as if the fish had measles. On occasion, we have seen individuals that were almost completely covered in these red spots. As their name implies, both when underwater and after capture, this species is capable of rapid color and pattern changes. On a number of occasions we have seen them quickly change color underwater, from scarlet to almost white. In addition, when first captured this species may be quite pale, only to turn uniformly pink or crimson soon after.

SIMILAR SPECIES
When we observe these fish underwater, and the dentigerous knobs and multiple-pointed lachrymal spines are not visible, this species is difficult to separate from splitnose and aurora rockfishes. However, splitnose have an inverted "U" pattern on their sides and no red spots on the lightly pigmented areas. Auroras have a light pigment pattern at about under the seventh through tenth dorsal spines, no red spots on the lightly pigmented areas, and their eyes do not rise above the cranium.

Note multiple points on lachrymal spine.

Sebastes melanostomus (Eigenmann & Eigenmann, 1890)
Blackgill Rockfish

Maximum Length: 61 cm (24 in) TL. **Geographic Range**: Northern British Columbia off west coast of Queen Charlotte Islands to Isla Cedros, central Baja California, typically from northern California to at least southern California. Pelagic juveniles have been taken as far south as Punta Abreojos (26°06'N, 114°05'W), southern Baja California strongly implying that adults live in southern Baja California. **Depth Range**: 88–768 m (288–2,520 ft), mostly in 200–600 m (656–1,968 ft). **Habitat**: This is a benthic species; we usually see them either as solitary individuals or in small groups. Most fish live in high-relief habitats, such as boulder fields and rock ridges. However, juveniles are often seen on soft substrates. **Description**: As adults, blackgill rockfish are heavy-bodied fish with a long and sloping head. *Juveniles*: Underwater, juveniles are pale pink or (occasionally) white with dark saddles that start in the dorsal fin and run down the sides, have darker blotches on the gill cover and pectoral fin base, white or pink tips at least on some fins, black in caudal and other fins, and a lateral line in a clear, pink zone. *Adults*: As the fish mature, they become mostly red, although dark or white bands, splotches, or saddles may be present, the clear lateral line becomes less apparent, and the fins may develop considerable black striations. In many individuals, the bands may run from the dorsal fin to the belly. In addition, the rear part of the gill cover is black (although this character is not always visible). The head of blackgill rockfish is quite large and the eye diameter about one-quarter that of the head length.

Juvenile.

SIMILAR SPECIES

No other northeastern Pacific rockfish has extensive dark pigmentation on the membrane at the rear of the gill cover. However, when this is not visible, blackgills may be most likely mistaken for darkblotcheds. Unlike blackgills, darkblotcheds have a series of five dark saddles on the back, these saddles just reach or barely dip below the lateral line (they often extend to the belly in blackgills), and the two blotches under the spiny dorsal fin often have merged to form a "W." Darkblotched rockfish also lack dark pigment on the gill membranes and have a more rounded snout. Splitnose rockfish lack pigment on the gill cover membrane, do not have five saddles, and have prominent dentigerous knobs. Shortraker rockfish have red saddles, but lack the pigment on the gill membranes.

Subadult blackgill rockfish (left) and splitnose rockfish.

Note black coloration.

Sebastes reedi (Westrheim & Tsuyuki, 1967)
Yellowmouth Rockfish

Maximum Length: 58 cm (23 in) TL. **Geographic Range**: Sitka, southeastern Gulf of Alaska to near San Francisco, northern California. **Depth Range**: 70–366 m (230–1,201 ft). **Habitat**: We know little about this species. It appears to live around complex habitats, such as rock ridges and boulder fields. **Description**: Yellowmouths are moderately deep-bodied with weak head spines and relatively small scales. We have not seen images of them underwater. After capture, they are primarily red or orange-red, with a dark blotch on the gill cover, vague to fairly prominent dusky saddles on the back, and yellow, black, or red blotches in the mouth and no dark edging on the fins. Individuals smaller than perhaps 40 cm (16 in) long have some black on their bodies and fins. **Note:** Declared "threatened" by the Committee on the Status of Endangered Wildlife in Canada.

SIMILAR SPECIES

As we have not seen this species underwater, we address identifying it after capture. No other northeastern Pacific rockfish species routinely has similarly colored blotches in the mouth (blackgill rockfish have dark areas at the rear part of the gill cover). Pacific ocean perch have a larger and sharper symphyseal knob and more prominent blotches on the back. Redstripe rockfish also have a larger symphyseal knob, are not as bright, tend to be slimmer, and have a light lateral line.

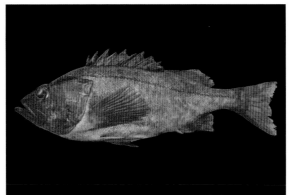

SCHON ACHESON

Sebastes crameri (Jordan, 1897)
Darkblotched Rockfish

Maximum Length: 59.5 cm (23.4 in) TL. **Geographic Range**: Eastern Bering Sea southeast of Zhemchug Canyon and Aleutian Islands off Tanaga Island to near Santa Catalina Island, southern California and Laguna Beach, southern California, most commonly from Yakutat, Gulf of Alaska to northern or perhaps central California. **Depth Range**: 29–915 m (95–2,985 ft). **Habitat**: This is a benthic species. Although we see them as solitary individuals, we have not observed enough of them to understand their habits. Both juveniles and adults are found mostly over mixed rock-mud sea floors; they will also venture out onto the mud and occasionally are found in high relief.

Description: This is a relatively deep-bodied rockfish. We have only rarely seen this fish underwater. *Juveniles*: Underwater, benthic juveniles are white with 4–5 wide, reddish-brown, or brown vertical bars that reach from the dorsal fins to the belly. *Adults*: In life, adults are basically white or pink with brown saddles (the remnants of the juvenile bars), often splashed with red. There are 4–5 saddles (caudal peduncle, soft dorsal, spiny dorsal (1–2), and nape). The saddle(s) under the spiny dorsal fin is usually incompletely constricted by a light area near its middle (forming a "W"); this separation is occasionally complete (forming two saddles). There is no light spot in the soft dorsal saddle, the fish have no obvious "<" mark behind the eye, but there is a dark blotch on the upper opercle. After death, adults have the same markings, although the body is pink or orange. **Note:** Declared "overfished" by the National Marine Fisheries Service and "threatened" by the Committee on the Status of Endangered Wildlife in Canada.

SIMILAR SPECIES

When viewed from certain angles (for instance, where the black on the rear of the gill cover is not visible), blackgill rockfish, particularly those individuals that are heavily blotched, may closely resemble this species. Both juvenile and adult blackgills are red or pink, with white or brownish blotching. Some of the dark blotching on blackgills, particularly that under the spiny dorsal fin, extends to the belly (rather than ending at about the lateral line as on darkblotcheds) and those under the spiny dorsal fin do not form a "W." In addition, darkblotcheds are more oval and perch-shaped.

ROBERT LAUTH

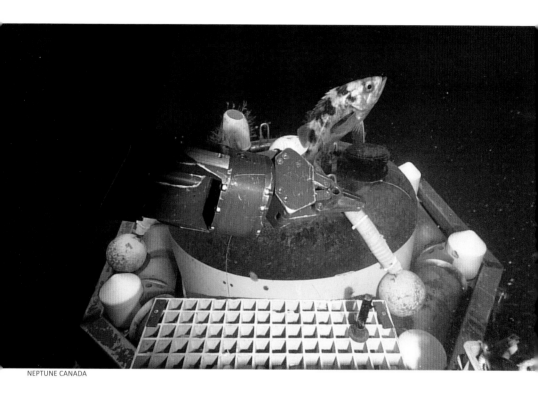
NEPTUNE CANADA

Sebastes borealis Barsukov, 1970
Shortraker Rockfish

Maximum Length: 120 cm (47.2 in) TL. **Geographic Range**: Okhotsk Sea, Pacific Ocean off northern Hokkaido, Japan to Kamchatka, and western Bering Sea to Navarin Canyon and Aleutian Islands to Point Conception, California; primarily from the northern Kuril Islands and southeastern Kamchatka Peninsula to Cape Olyutorskiy in the western Bering Sea and at least to British Columbia. **Depth Range**: 25–1,200 m (82–3,937 ft), mostly at perhaps 300–600 m (984–1,640 ft). **Habitat**: These are bottom-oriented and generally solitary fish, although they will form small aggregations. They live over rocks and either rest on the bottom or hover as much as 10 m (33 ft) in the water column. **Description**: Shortrakers are robust and heavy-bodied fish. As their name implies, this species has very short, knob-tipped gill rakers. Dead individuals tend to be pink or red and lack much in the way of markings, although the fins are often black edged. We have reviewed only a few images of this species in its natural habitat. However, it appears that, underwater, both juveniles and adults are mixtures of red, pink, or pink-orange, with white or light saddles and blotches, particularly on the head and back. There is often white edging on the fins.

SIMILAR SPECIES
When viewed underwater, blackgill, blackspotted, and rougheye rockfishes might be confused with this species. Blackgills have considerably narrower heads (they are generally less robust), are usually bright red (occasionally lighter red), the rear of the gill cover is black, and there are usually some black striations on at least some of the fins. Blackspotteds are variously pink to red, sometimes with a greenish to black wash, and always with dark spots and mottling over much of the body. Rougheyes are red, pink, or reddish-orange, with faint bands of brownish-red mottling on the flanks and dark blotches on the head.

GREENPEACE

Probably an adult on the right and perhaps a juvenile on the left.

Sebastes aleutianus (Jordan & Evermann, 1898)
Rougheye Rockfish

Also in Asia

Maximum Length: 72.6 cm (28.6 in) TL. **Geographic Range**: North Pacific from at least the eastern Aleutian Islands off Unalaska Island, the eastern Bering Sea at 59°11'N, south to southern Oregon at 43°54'N, and perhaps into northern California. **Depth Range**: 45–765 m (158–2,509 ft). **Habitat**: Rougheyes live over rocks and adjacent soft sea floors. **Description**: A complete description of this species is found in Orr and Hawkins (2008). Rougheyes are relatively heavy-bodied fish, with a deep and robust head and numerous head spines (eight pairs of strong head spines and two or more spines below the orbit, unique, other than in blackspotteds, to this species). They are red, pink, or reddish-orange, with faint bands of brownish-red mottling on the flanks. Dark blotches are found on the head, particularly on the gill cover. While the body mottling is sometimes quite heavy, there are no distinct spots. **Note:** Declared "threatened" by the Committee on the Status of Endangered Wildlife in Canada.

SIMILAR SPECIES

Blackspotted rockfish are most likely to be confused with rougheyes, although in most cases, both after capture and underwater, distinguishing rougheyes from blackspotted rockfish is a straightforward proposition. Fish that are light-colored and heavily spotted or dark-colored and lightly spotted are blackspotteds. Confusion can occur in the small percentage of rougheyes that have diffuse blotches at the base of the dorsal fin (these may appear to be spots) or those blackspotteds that are light-colored with only one or two spots (and occasionally none). Jay Orr informs us that in the laboratory a very small percentage of rougheyes cannot be distinguished from blackspotteds, using color, pattern, meristics, or morphometrics. At least some of these individuals are hybrids of the two species. In many instances, comparing the length of the first dorsal spine to the orbit length is a useful character, as blackspotteds have a longer spine (1.0–1.8 when divided into the orbit length) than do rougheyes (1.5–3.0 into orbit length). Rougheyes may also be confused with shortraker and blackgill rockfishes. Underwater, shortrakers tend to be blotchy reddish-orange and white, often with white-tipped dorsal spines. After capture, they are readily separated from rougheyes by their short, knobby-tipped gill rakers on the first arch. Blackgills are usually bright red, with a relatively clear lateral line and often with quite distinct white or brown saddles and blotches. There is a black rim around the bottom and posterior edges of the gill cover.

NOTE

Because of previous confusion with blackspotted rockfish, the maximum size, geographic and depth ranges of this species may not be well understood.

Sebastes melanostictus (Matsubara, 1934)
Blackspotted Rockfish

Also in Asia

Maximum Length: To at least 60 cm (23.6 in) TL.
Geographic Range: Western Japan (approx. 36°N, 141°E) north to the Bering Sea, (61°49'N, 177°20'W) to San Diego, California (32°35'N, 117°24'W). **Depth Range**: 84–1,000 m (276–3,280 ft). **Habitat**: Blackspotteds are benthic and near-bottom fish that live over high relief and nearby soft sea floors. **Description**: A complete description of this species is found in Orr and Hawkins (2008). Blackspotteds are heavy bodied and spiny headed (with a complement of eight pairs of strong head spines and two or more spines below the orbit, unique, other than in rougheyes, to this species). Orr and Hawkins report that some blackspotted rockfish are "pink to red with greenish black spotting above and below the lateral line, often extending on to the spinous and soft dorsal fin." However, there are also dark individuals "with black to greenish wash over red to pink background color, heavily mottled over entire body and spotted at and above the lateral line; heavy dark mottling often obscuring spotting. Body rarely dusky overall but having diffuse mottling, without spotting." **Note:** Declared "threatened" by the Committee on the Status of Endangered Wildlife in Canada.

PAUL SPENCER

Juvenile.

SIMILAR SPECIES

As we have noted in the rougheye rockfish species account, due to the considerable variability in key color and pattern characteristics (rarely due to hybridization), some individuals of the blackspotted-rougheye complex are not easily assigned to species, particular those observed underwater. As noted above, a majority of blackspotteds are either heavily spotted or, if lightly spotted, are dark-colored, darker than rougheyes. Rougheye rockfish are pink, red, or reddish-orange and lack spots on the body or fin membranes. However, some individual blackspotted are both light-colored (as are rougheyes) and have either few spots or (in rare instances) none at all. In addition, there is the occasional rougheye that has diffuse blotching at the base of the dorsal fin and this can be confused with spotting. In some instances, this makes accurate identification on ship board, and particularly underwater, difficult or impossible. Jay Orr reports that comparing the length of the first dorsal spine to that of the orbit can be a useful tool in differentiating problem individuals. The first spine in blackspotteds is longer (1.0–1.8 when divided into the orbit length) than are those of rougheyes (1.5–3.0 into orbit length). Given sufficiently sharp images, this technique might also be effective with images of fish taken underwater. Blackgill rockfish may live in the same waters as blackspotteds. Blackgills are usually an intense red with a lighter lateral line, often (but not always) with white head, dorsal, and/or lateral blotches, lack dark spots, and have black pigment on the gill cover membrane. Shortraker rockfish are blotchy reddish or orange and white; they, too, lack dark spots.

NOTE

Because of previous confusion with rougheye rockfish, the geographic and depth ranges and maximum size of this species are poorly understood.

All of these are blackspotted rockfish.

Blackspotted on the left and blackgill on the right.

Sebastes pinniger (Gill, 1864)
Canary Rockfish

Maximum Length: 76 cm (30 in) TL. **Geographic Range**: Pribilof Islands, eastern Bering Sea, and western Gulf of Alaska south of Shelikof Strait to Punta Colnett, northern Baja California, typically from at least as far northward as British Columbia to central California. **Depth Range**: Young fish in shallow waters; adults about 18–838 m (59–2,749 ft) and mostly from 80–200 m (262–656 ft). **Habitat**: This is a schooling species that, while usually found near the bottom, will ascend well up into the water column. Young fish settle from the plankton to near the bottom (often in shell hash), in inshore waters, often along the edges of rocks and kelp and rarely in tide pools. These juveniles move into somewhat deeper waters within a few months. Older juveniles and adults primarily inhabit high-relief structures, such as rock ridges and boulders, but also dwell on cobble and on mud-rock habitats. **Description**: Canary rockfish are spiny and moderately deep-bodied fish. Juveniles are slimmer with relatively large heads. *Juveniles*: When they first settle from the plankton, juveniles are pale, with a large black spot on the rear of the spiny dorsal fin, and brownish saddles on back and sides. Fish as large as 20 cm (6 in) may retain remnants of the black spot. By about 4 cm (1.6 in), these fish have become brassy, several eye stripes have developed, and the lateral line is clear. *Older Juveniles and Adults*: Underwater, larger juveniles and adults are yellow or orange across a gray or white background, with a distinct, white or gray lateral line that runs from the head to the caudal peduncle (this becomes fainter or turns orange after death), three diagonal stripes across the head (including one on either side of the eye), and the leading edges of the pelvic and anal fins are white. The pelvic and anal fins have a very pointed and angular outline. **Note:** Declared "overfished" by the National Marine Fisheries Service and "threatened" by the Committee on the Status of Endangered Wildlife in Canada.

Young-of-the-year.

Juvenile.

SIMILAR SPECIES
Vermilion rockfish Type 1 are red (sometimes with black markings), rather than the orange or yellow of canaries, and do not have the white-edge anal fin. Their pelvic fins are rounded rather than angular. Vermilion rockfish Type 2 do not have gray markings and their caudal fins tend to be darker than the rest of the body. These two species also have a clear lateral line, but it does not reach all the way to the head as it does in canaries. Yelloweye rockfish have a bright yellow eye, lack the head striping, do not have the gray or white background, and, in general, have a much craggier appearance.

MARC CHAMBERLAIN CSUMB MARE

SWFSC ROV TEAM

TOM LAIDIG

Sebastes miniatus (Jordan & Gilbert, 1880)
Vermilion Rockfish

Note: Studies by John Hyde demonstrate that the "vermilion rockfish" is composed of two species, here called Type 1 and Type 2. While the juveniles of both species appear to live in shallow waters, adults appear to be somewhat separated by depth, with adult Type 1 rockfish tending to occupy waters less than about 100 m (328 ft) and adult Type 2 living in waters deeper than that depth. However, as the maximum length, and geographic and depth ranges have not been clearly defined for the two species, the data we present represents an aggregate of the two species.

Maximum Length: 76 cm (30 in) TL. **Geographic Range**: Zaikov Bay, Montague Island, Alaska (approx. 60°17'N, 147°04'W) to Islas San Benito, Baja California (28°18'N, 115°35'W) and Isla Guadalupe, central Baja California. **Depth Range**: 6–478 m (19–1,568 ft). **Habitat**: Cobble, boulders, low and high relief rocks, occasionally over sand and mud. **Description**: *Juveniles*: It is likely that newly settled young-of-the-year of both species are mottled brown with black patches in the spiny and soft dorsal fins and a clear caudal fin. Older juveniles are mottled orange or red.

Adults: Type 1 fish are brick-red, sometimes with black or gray markings, while Type 2 rockfish are yellowish-orange, although the fins may be red. We believe that the caudal, anal, and paired fins of Type 2 rockfish tend to be darker than the body; this is not the case with Type 1 rockfish. Compared to Type 1 rockfish, Type 2 rockfish reportedly have smaller eyes, are wider between the eyes, and have a more slender caudal peduncle. While we are comfortable identifying to species most of the Type 1 and Type 2 fish we observe underwater, we find that we have difficulty with some individuals.

Young-of-the-year.

Juvenile.

SIMILAR SPECIES
Canary rockfish are orange or yellow, and have a clear lateral line that extends from the head to the base of the caudal fin. The lateral line in Type 1 and Type 2 rockfishes is also clear, but does not extend on to the head. Yelloweye rockfish have a bright yellow eye, lack the head striping, do not have the gray or white background, and, in general, have a much craggier appearance.

Type 1 rockfish.

Type 1 rockfish.

Type 2 rockfish.

Type 2 rockfish.

Sebastes ruberrimus (Cramer, 1895)
Yelloweye Rockfish

Maximum Length: 91.4 cm (36 in) TL and perhaps to 103.5 cm (40.8 in) TL. **Geographic Range:** South of Umnak Island (Alaska) (53°17'N, 168°22'W) to Ensenada, northern Baja California, and perhaps to Bahia San Quintin, northern Baja California, and common from perhaps the western end of the Alaska Peninsula to central California. **Depth Range:** 11–549 m (36–1,800 ft), typically from 91–180 m (300–590 ft). Adults tend to live in shallower water in the north. **Habitat:** This is a species primarily of complex, high relief habitats, although near Kodiak Island we observed a number of them on flat, sandy bottom containing isolated, low-relief rocks. It primarily lives on or near the bottom. Juveniles are found in nearshore areas. Adults are territorial and spend much of their time sheltering in crevices, although they will often venture a few meters above the sea floor. While we usually see them as solitary individuals, Lynne Yamanaka and Rick Stanley surveyed fishes at Bowie Seamount (British Columbia), a less-heavily fished site well off the coast, and observed aggregations of 30 or more adults a short distance off the bottom. **Description:** Yelloweye rockfish are heavy-bodied and spiny fish. A key character is the bright yellow eye found in fish larger than about 25 cm (10 in). *Juveniles and Subadults:* Newly settled young-of-the-year may be very dark. In addition, as noted by O'Connell (2002) "Newly settled juveniles [young-of-the-year] are dark red-orange and have two horizontal bright white stripes, one on each side of the lateral line. The fins of these juveniles may be fringed with either black or white and there is usually a white vertical band at the insertion of the caudal peduncle, along with white patches along the back and at the base of the dorsal fin. As the fish matures, the stripes fade and the color changes from red-orange to a lighter orange. Subadults lose the lower white stripe and often have white fins." *Adults:* These are dark brown-red, red-orange, or orange and may have lighter blotches on back and dorsal fins. Individuals larger than 30 cm (12 in) have a raspy ridge on their heads. **Note:** Declared "overfished" by the National Marine Fisheries Service and "threatened" by the Committee on the Status of Endangered Wildlife in Canada.

SIMILAR SPECIES
The most likely species to mistake for the yelloweye are the canary, and Type 1 and Type 2 vermilion rockfishes, as all have somewhat similar body shapes and colorations. However, none of these species has a bright yellow eye.

Newly settled young-of-the-year.

Somewhat older juvenile.

A typical subadult.

A darker subadult.

Note lighter blotches on back and fins.

A number of yelloweyes at Bowie Seamount are dark with distinctive gray markings on the head.

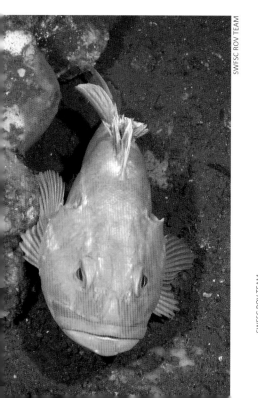

Sebastes levis (Eigenmann & Eigenmann, 1889)
Cowcod

Maximum Length: 100 cm (39.4 in) TL. **Geographic Range**: Northern Oregon (44°44'N, 124°40'W) to Banco Ranger (28°36'N, 115°53'W) and Isla Guadalupe, central Baja California, primarily from central California to at least off Bahia San Quintin, northern Baja California. **Depth Range**: Juveniles as shallow as 40 m (132 ft); adults 72–491 m (236–1,690 ft), most adults at about 130–215 m (426–705 ft). **Habitat**: This is a benthic species. Juveniles mostly settle out on such hard and low relief as cobble, small boulders, pipelines, and the shell mounds that surround oil platforms. Older juveniles and adults live in high and complex relief (e.g., rock ridges, boulder fields, and the bottom of oil platforms). **Description**: Cowcod are one of the more easily recognized of the northeastern Pacific rockfishes. As adults, they are deep-bodied and have a very high, and deeply incised, spiny dorsal fin. *Juveniles*: Benthic young-of-the-year are cream, white, or pink with irregular, and often broken, dark or red vertical bars. *Older Juveniles and Adults*: The bars of young-of-the-year tend to fade with age, but even large adults may exhibit faint remnants. Older juveniles and adults range in color from cream through pink, salmon, orange, and golden. Juveniles and small adults have darker bars radiating from the eye. **Note:** Declared "overfished" by the National Marine Fisheries Service.

Young-of-the-year.   SWFSC ROV TEAM

SIMILAR SPECIES

Juvenile cowcod might be confused with the several other vertically barred species, such as tiger, flag, and redbanded rockfishes, and treefish. However, whereas the bars of young cowcod are somewhat amorphous and clearly composed of a series of spots or small blotches, those of the former species are solid. In addition, the coloration is different in young cowcod, as juvenile tigers are red with dark vertical bars, both flag and redbanded rockfish are white with red or orange bars, and treefish are black or dark green alternating with silvery or yellow bars. In the older juvenile and adult stages, bronzespotted rockfish might be mistaken for cowcod. However, bronzespotteds have a distinctly upturned mouth, a low and not deeply incised dorsal fin, large dark spots on body and fins, and usually are white blotched.

Juvenile.   SWFSC ROV TEAM

Subadult.

Sebastes gilli (Eigenmann, 1891)
Bronzespotted Rockfish

Maximum Length: 85 cm (33.5 in) TL. **Geographic Range**: West coast of Vancouver Island (about 49°N, 127°W), British Columbia to Punta Colnett (30°53'N, 116°30'W), northern Baja California, most often from southern California southward. **Depth Range**: 75–413 m (246–1,362 ft), mostly found between 130–250 m (426–820 ft). **Habitat**: Bronzespotteds are benthic fish that live around high-relief rocks and amid boulders. We see so few of them that it is unclear if this is a solitary species or, in areas of high density, an aggregating one. **Description**: Bronzespotted rockfish could have patented the term "squat." These fish are quite deep-bodied and spiny, have a relatively low and not deeply incised spiny dorsal fin, and a very distinctive sharply upturned mouth. Juveniles share the morphology, color, and color patterns of the adults, although they are not as deep-bodied. The base color of this species ranges from red and orange-brown to yellow with large brown spots on the body and fins. Underwater, the head, back, and sides often have white patches (these tend to disappear in dead specimens) and a series of darker stripes radiate backwards (and some downwards) from the upper jaw and eye. Some or all of the fins may have black striations.

Juvenile.

SIMILAR SPECIES
Of all of the northeastern Pacific species, only the cowcod might be easily mistaken for bronzespotteds. However, cowcods do not have an upturned mouth, do have a high and deeply incised spiny dorsal fin, and lack dark spots. Bank rockfish lack the strongly upturned jaw, tend to have smaller spots, and are more ovate and less squat.

SWFSC ROV TEAM SWFSC ROV TEAM

SWFSC ROV TEAM SWFSC ROV TEAM

SWFSC ROV TEAM

Sebastes rosenblatti Chen, 1971
Greenblotched Rockfish

Maximum Length: 54 cm (21.3 in) TL. **Geographic Range**: Point Delgada, northern California (40°01'N, 124°04'W) to Isla Cedros (26°06'N, 115°20'W) and Isla Guadalupe, central Baja California and typically from central California to Banco Ranger and Isla Guadalupe, central Baja California. **Depth Range**: 55–491 m (182–1,610 ft), in southern California, mostly in 170–270 m (558–886 ft). **Habitat**: This is a solitary or semi-solitary species that usually sits right on, or hovers barely above, the bottom. They live on rocks, boulders, and mixed rock-mud bottoms; the bottoms of several oil platforms have high densities of them.

Juvenile.

Description: Greenblotched rockfish are squat and spiny fish with a deeply incised spiny dorsal fin. Their body colors range from reddish, yellow, and orange to almost white. Key characters are the distinct green or green-brown vermiculations above the lateral line and on top of the head (confusingly there may also be some distinct spots), and a spiny dorsal fin that is heavily striated with green and salmon or orange. Larger individuals develop spinier and craggier heads.

SIMILAR SPECIES
Underwater, particularly when visibility is poor, we occasionally have considerable problems differentiating greenblotcheds from greenspotted rockfish, as the diagnostic spotting or vermiculations may not be easily seen. However, the spiny dorsal fin of a typical greenspotted is primarily white or very light, with only a wash of green or pink, compared to the extensive green and salmon striations of the greenblotched. Underwater, pink rockfish may also be confused with greenblotcheds. However, pinks do not have green vermiculations on back and head; rather their back is "smudged" with green or brown. When visibility allows for clear photographs or frame grabs, the usually 17 pectoral rays of the greenblotched separates it from the primarily 18-rayed pink. Differentiating large pink and greenblotched rockfishes presents a problem that we have sometimes not been able to solve, as both species have craggy features, similarly colored fins, and the vermiculations on very large greenblotcheds may become faint. Freckled rockfish have a distinct green or brown reticulated pattern on the back and dentigerous knobs on the upper jaw.

Note the 17 pectoral rays. The vermiculations on older fish often fade.

77

Sebastes chlorostictus (Jordan & Gilbert, 1880)
Greenspotted Rockfish

Maximum Length: 53.4 cm (21 in) TL. **Geographic Range**: Vancouver Island (49°04'N, 126°50'W) to southern Baja California (25°32'N, 113°04'W) and Isla Guadalupe, central Baja California. They are common from central California to at least off Bahia San Quintin, northern Baja California. **Depth Range**: 30–379 m (100–1,243 ft), mostly in 80–160 m (262–525 ft). **Habitat**: Greenspotteds are benthic and mostly solitary fish, although you will occasionally see a few lodged next to each other in a crevice. They live primarily over mixed low-high relief habitat, but will venture out over sand and mud.

Description: This is a compact, heavy-bodied, and spiny species with a large and deeply incised spiny dorsal fin. Underwater, juveniles are light or pink, while adults range from orange to pink and varieties of yellow or gold. From the time they settle out of the plankton, this species is profusely marked with green spots (these range from quite large to very small) above the lateral line and on top of the head. Characteristically, the spiny dorsal fin is either white, white with a greenish tinge, or pink. There are 3–5 white or pinkish blotches on the back.

SIMILAR SPECIES

Underwater, the closely related greenblotched and pink rockfishes are the species most likely to be confused with greenspotteds, as both have very similar morphologies and both have markings on the back. When water conditions permit, the green vermiculations of greenblotcheds readily separate this species from the greenspotted. However, in dark or turbid waters or when seen from a distance, where vermiculations are difficult to

Juvenile.

see, separating these two species is more problematic. One feature that we believe is most helpful is that the spiny dorsal fin of greenblotcheds is always heavily pigmented with green and, secondarily, orange, while that of the greenspotted is mostly white or white washed with green. Pink rockfish lack the distinct green spots and have green and pink dorsal spines. Larger greenblotcheds and pinks also have much craggier heads.

SWFSC ROV TEAM

SWFSC ROV TEAM

SWFSC ROV TEAM

SWFSC ROV TEAM

79

Sebastes eos (Eigenmann & Eigenmann, 1890)
Pink Rockfish

Maximum Length: 56 cm (22 in) TL. **Geographic Range**: Oregon (44°33'N) to southern Baja California (25°24'N, 113°01'W) and Isla Guadalupe, central Baja California, primarily from southern California southward. **Depth Range**: 45–366 m (150–1,200 ft), commonly 200–350 m (656–1,148 ft). **Habitat**: Pinks are benthic, and usually solitary, fish. We usually see them in high- and low-relief rocky areas.
Description: Pink rockfish are deep–bodied spiny rockfish with, usually, five white spots on the side and back. We believe that the front of the head of this species is often quite steep and blunt. Often, the soft dorsal, caudal, and anal fins are outlined in white or lighter coloration. Underwater, this species is usually white, pink, or brassy, with diffuse greenish or brown marks or vermiculations on the back. In larger individuals, these markings appear particularly amorphous (often as just a faint dusting) and "smeared." Characteristically, larger fish have very large and craggy heads. Although not observable underwater, the gill rakers are particularly short and stubby.

SIMILAR SPECIES
We are faintly uncomfortable with the following discussion, as pinks are most likely to be confused with greenblotcheds and there have been a number of times we were unable to distinguish these two species when viewed underwater. Secondarily, greenspotteds may also cause problems. Pinks often (but not always) lack distinct green or green-brown vermiculations (always found in greenblotcheds) and do not have the myriad green spots of greenspotteds. In addition, pinks usually have 18 pectoral rays (compared to 17 in greenblotcheds). We think that greenblotcheds tend have a less blunt (somewhat more elongate) head. Lastly, larger pink rockfish have much craggier heads than greenspotted rockfish.

SWFSC ROV TEAM

SWFSC ROV TEAM

SWFSC ROV TEAM   SWFSC ROV TEAM

SWFSC ROV TEAM

We think the lower fish is a pink because it has 18 pectoral rays and no spots or vermiculations on its back. The other fish is a bank rockfish.

SWFSC ROV TEAM

81

Sebastes ensifer Chen, 1971
Swordspine Rockfish

Maximum Length: 30.5 cm (12 in) TL. **Geographic Range**: San Francisco, northern California to Banco Ranger (28°25'N, 115°32'W), central Baja California and Isla Guadalupe, central Baja California, and abundant from the Santa Barbara Channel, southern California southward. **Depth Range**: 50–433 m (164–1,420 ft), primarily in 90–240 m (295–787 ft). **Habitat**: Swordspines are benthic and while they do not form aggregations, we often see several of them sheltering within the same rock crevice. **Description**: Swordspine rockfish are slender with 3–5 white or light blotches on the back, a second anal spine longer than the soft anal rays, and a slightly projecting lower jaw with a relatively strong symphyseal knob. Underwater, they are often lightly striped and range from pale pink through yellow-orange to red, with a yellow wash. Both the top of the head and the body may have extensive white lines or webbing (a single white patch encircled in pigment is often visible on the top of the head between the eyes). An excellent, and key, character is that the long second anal spine is sheathed in white and is often visible in low light or turbid waters. On deck, swordspines become pink or crimson with some greenish-yellow on the back.

SIMILAR SPECIES
Underwater, pinkrose and rosethorn rockfishes pose the greatest potential for confusion, as they closely resemble swordspines and live in similar habitats over a broad depth range. Pinkroses are deeper bodied, do not have a slightly projecting lower jaw, and, most importantly, none of their anal spines are longer than the anal soft rays. As with swordspines, rosethorns are also striped, but they are deeper bodied and their anal spines are not longer than (although they can be as long as) the soft anal rays. Rosy rockfish are mostly purple and red-orange, somewhat deeper bodied, do not have a long second anal spine, and have purple on the head and around the white blotches

Note white-sheathed second anal spine.

Sebastes simulator Chen, 1971
Pinkrose Rockfish

Maximum Length: 42.1 cm (16.4 in) TL. **Geographic Range**: At least Carmel Submarine Canyon, central California and perhaps to off Eureka, northern California to Cabo Colnett (30°53'N, 116°30'W), northern Baja California and Isla Guadalupe, central Baja California, perhaps most commonly from southern California southward. **Depth Range**: 99–450 m (325–1,476 ft). Off southern California, they are mostly in 205–320 m (672–1,050 ft). **Habitat**: Pinkrose are benthic and primarily solitary fish (although several will often huddle near each other in a crevice). This species lives in either high- and low-relief rocks and other complex structures. **Description**: These are deep–bodied, wide-headed, and spiny rockfish with 3–5 large white blotches (and often several other smaller ones) on the back. Their body color ranges from cherry-red to pink-red and may be washed with yellow or have some vague greenish marks on back and dorsal fins, but they do *not* have body striping. White webbing may occur on the top and sides of the head and on the anterior flanks. Many individuals have a somewhat darker "<" mark behind each eye. The second anal spine is stout and longer than the third, but not longer than the anal rays.

SWFSC ROV TEAM

SIMILAR SPECIES

When viewed underwater, we have considerable difficulty differentiating pinkrose from rosethorn rockfish and we do not believe we have developed a foolproof method for doing so. Fish that are bright red or reddish with a yellow flush, but lack striping, with 17 or more pectoral rays, are almost certainly pinkroses. Fish with 16 pectoral rays, rather than 17, and with reddish, green, or yellow striping are likely rosethorns. In certain circumstances, distinguishing swordspine rockfish from pinkroses also poses a problem. This is particularly the case where both species are in crevices (and thus the anal spine length cannot be determined) and when both species are similarly a pink-yellow. Rosy rockfish are somewhat similarly patterned, however they have a considerable amount of purple on head, back, and sides.

SWFSC ROV TEAM

Sebastes helvomaculatus Ayres, 1859
Rosethorn Rockfish

Maximum Length: 43 cm (6.9 in) TL. **Geographic Range**: Western Gulf of Alaska east of Sitkinak Island to Banco Ranger (28°33'N, 115°25'W), central Baja California, commonly from about the Yakutat, Gulf of Alaska, area to central California. **Depth Range**: 16–1,145 m (52–3,756 ft), typically in 80–350 m (262–1,148 ft). **Habitat**: Rosethorns are benthic fish that are most often found on mixed rock-mud bottoms, but will also occupy boulder fields and muddy sea floors. While basically solitary animals, several individuals will sometimes shelter under the same ledge. **Description**: Rosethorn rockfish are small, spiny rockfish with 4–5 white blotches on the back, anal spines shorter than or equal to the soft anal rays, and a relatively narrow head. Underwater, they are a mixture of pale pink, yellowish-green, yellow, orange, or crimson, arranged in a series of diffuse stripes, and they often have green on the top of the head. The dorsal and anal fins are usually striped with green or pink and the head may have white streaks.

SIMILAR SPECIES
When viewed underwater, we often have great difficulty separating pinkrose rockfish from rosethorns. We believe that body color probably provides one of the better differentiating characters, although it is not perfect. In particular, it is our sense that if an individual has the green, yellow, or red striping it is a rosethorn. Additionally, if still images allow, counting pectoral rays is also useful, as pinkroses usually have 17 rays, while most rosethorns have 16 rays. Swordspine rockfish are slimmer and have a second anal spine that is longer than the soft anal fin. Rosy rockfish are yellow to red, usually with purple waves and vermiculations on top of the head and back and with the lighter blotches on the back rimmed in purple.

Sebastes rosaceus Girard, 1854
Rosy Rockfish

Maximum Length: 36 cm (14 in) TL. **Geographic Range**: Strait of Juan de Fuca, Washington (48°24'N, 124°43'W) to Bahia Tortugas (27°40'N, 114°52'W) and Isla Guadalupe, central Baja California typically from at least Cordell Bank, northern California to at least Punta Colnett, northern Baja California, and on Banco Ranger, central Baja California. **Depth Range**: 7–328 m (24–1,076 ft), in southern California most commonly in 40–100 m (131–328 ft). **Habitat**: Juveniles and adults live mostly over high-relief bottoms, as well as over cobble and occasionally on sand and mud near high relief. These are mostly solitary fish, although they will form little groups on occasion. They tend to either lie right on the sea floor or hover a meter or two above it. **Description**: Rosy rockfish are diminutive, spiny fish with 4–6 light blotches on the back. Underwater, rosies are yellow-orange, orange, or red, and heavily marked and striated with purple on head and back. Characteristically, these blotches are outlined in purple or red. The purple on the head and around the light blotches may fade to pink in some fish. Confusingly, an occasional individual will show very heavy white markings and webbing on dorsal and caudal fins, back, and sides.

Juvenile.

SIMILAR SPECIES
Swordspine and, perhaps, pinkrose and rosethorn rockfishes, diminutive species with similar morphologies and light dorsal blotches, may be confused with rosies. Swordspines are slimmer, are pink through yellow-orange and red (with no purple), and have that very long second anal spine. Similarly, pinkroses (pink or red with some yellow on the sides and in the fins) and rosethorns (diffusely striped pale pink, yellowish-green, yellow, orange, or crimson) lack the rosy's purple body coloration.

We see this pattern only occasionally.

Sebastes lentiginosus Chen, 1971
Freckled Rockfish

Maximum Length: 23 cm (9 in) TL. **Geographic Range**: Point Conception (34°34'N, 120°37'W), southern California to southern Baja California (25°61'N, 113°24'W), but rarely northward of Santa Monica Bay, southern California. **Depth Range**: 22–290 m (73–951 ft). **Habitat**: This is a benthic species and we see them over both high and low relief reefs, either as solitary individuals or perhaps two to three fish lying near each other. **Description**: Freckled rockfish are small, compact rockfish with two very distinctive and pointed dentigerous knobs on the front of the upper jaw (one on each side of the snout). The species' base colors can range from pink-brown to orange-brown with five or more lighter blotches on the back and sides. Of particular importance, the back (primarily above the lateral line) is heavily marked with broken striations or freckling of green or greenish-brown and the sides below the lateral line can be either lightly cross-hatched or without much distinct markings.

SIMILAR SPECIES

Particularly underwater, honeycomb rockfish may look very similar to freckled rockfish. However, honeycombs lack the very distinct dentigerous knobs, the third anal spine is shorter than the anal rays (longer than or equal to in freckleds), the dark markings on the back form amorphous "smudges" rather than distinct striations or streaks, and the species has a honeycomb pattern on the sides (caused by brown or green edging to the scales). We note, however, that this latter pattern may not be visible under low-light conditions. Among the other closely related rockfishes, only the greenblotched might cause confusion. Greenblotcheds are much deeper bodied, lack the distinct dentigerous knobs, have a larger and higher spiny dorsal fin, and their vermiculations form a less organized pattern. In addition, the anal spine of a freckled rockfish is almost the same length as or longer than the anal rays; it is half that length in greenblotcheds.

This is either a juvenile freckled rockfish or honeycomb rockfish. TERRY STRAIT

SWFSC ROV TEAM

SWFSC ROV TEAM

SWFSC ROV TEAM

Sebastes umbrosus (Jordan & Gilbert, 1882)
Honeycomb Rockfish

Maximum Length: 28.5 cm (11.2 in) TL. **Geographic Range**: Point Pinos, central California (36°38'N, 121°56'W) to Bahia San Juanico (26°15'N, 112°28'W), and Rocas Alijos, southern Baja California, primarily from about Santa Monica Bay, southern California, southwards. **Depth Range**: 18–270 m (60–891 ft), perhaps typically between 30–90 m (98–295 ft). **Habitat**: Both juveniles and adults live mostly over boulders or cobble. These are solitary fish that most often shelter in crevices, but will also ascend a few meters above the sea floor. **Description**: Honeycomb rockfish are compact, squat, and spiny rockfish with 4–6 light blotches on their back. Their base body colors are tan, brown, reddish-brown, or pink-yellow and, below the lateral line, each scale is edged with dark brown or green, producing the characteristic honeycomb pattern. In addition, their backs have brownish or greenish-brown patches or smudging. Occasionally, we see a color and pattern variant that is brown and almost white, washed with pink, and whose blotches are dark pink. These individuals have reduced honeycombing.

SWFSC ROV TEAM

SIMILAR SPECIES

Freckled rockfish can be easily confused with honeycombs as they have a somewhat similar body shape, color, and pattern. However, older juveniles and adult freckleds have two prominent dentigerous knobs on their upper jaws (one on either side of the snout) and very distinct green or brown streaks, striations, or freckling on the back, mostly above the lateral line (compared to the "smeared" mottling on honeycombs). In addition, they lack the heavy honeycomb pattern on the flanks.

SWFSC ROV TEAM

SWFSC ROV TEAM

SWFSC ROV TEAM

SWFSC ROV TEAM

This is the unusual color pattern.

SWFSC ROV TEAM

93

Sebastes notius Chen, 1971
Guadalupe Rockfish

Maximum Length: 21.9 cm (8.6 in) SL. **Geographic Range**: This species has been collected at two sites: Isla Guadalupe, central Baja California and in the vicinity of Banco del Tio Sam (Uncle Sam Bank; 25°35'N), southern Baja California. **Depth Range**: 165–250 m (541–820 ft). **Habitat**: Likely rocky sea floors. **Description**: This is a small, compact and spiny fish, characterized by 2–5 lighter blotches on the back. After capture, they are orange-red or yellow, with red vermiculations on the back and top of head. We do not know what color they are underwater.

SIMILAR SPECIES
This is likely the least known of the northeastern Pacific rockfishes, as only a handful of specimens have been caught and, as far as we know, none have been observed underwater. After capture, Guadalupe rockfish most closely resemble swordspines. However, swordspines are slimmer and the second anal spine is longer than the anal soft rays (it is shorter in Guadalupe rockfish).

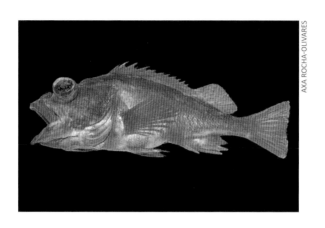

Sebastes constellatus (Jordan & Gilbert, 1880)
Starry Rockfish

Maximum Length: 46 cm (18 in) TL. **Geographic Range**: Off the mouth of Russian River (38°27'N, 123°07'W), northern California to off Todos Santos (23°24'N, 110°14'W), southern Baja California and Isla Guadalupe, central Baja California. Typically from central California to at least off Bahia Tortugas, southern Baja California. Tentatively identified from photographs taken at Rocas Alijos, southern Baja California. **Depth Range**: 15–274 m (50–900 ft), primarily in 40–160 m (131–525 ft). **Habitat**: Benthic, and generally solitary, starries are found over high-relief rock ridges, boulders, and cobble. **Description**: Starry rockfish are very attractive and easily identified fish. *Juveniles*: Unique among northeastern Pacific rockfishes, newly settled juveniles are lemon-yellow. *Older Juveniles and Adults*: At about 10–15 cm (4–6 in), juveniles are pink and the spots begin to appear. These fish are spiny and somewhat elongated, with a rounded and short head, 4–5 lighter blotches on the back, and orange, red, or orange-brown skin that is covered in numerous small white, yellow, or (after capture) blue spots.

SIMILAR SPECIES
It is unlikely that any other northeastern Pacific rockfish will be mistaken for this species, as the combination of intense spotting, orange or red coloration, the 4–5 lighter dorsal patches, and the characteristically rounded head make this fish quite distinctive. Whitespeckled rockfish (see below) also have light spots, but these fish are slimmer and lack the 4–5 light blotches on the back.

Juvenile.

Starry rockfish (top) and whitespeckled rockfish.

SWFSC ROV TEAM   SWFSC ROV TEAM

Sebastes moseri Eitner, Kimbrell, & Vetter, 1999
Whitespeckled Rockfish

Maximum Length: 20.6 cm (8.2 in) TL. **Geographic Range**: Off Big Creek, central California (36°03'N, 121°36'W), central California to Punta Colnett, Baja California (30°57'N, 116°20'W), abundant northward as far as the southern part of southern California. **Depth Range**: 50–274 m (164–900 ft), mostly commonly between 80–200 m (262–656 ft). **Habitat**: Whitespeckleds live near the bottom, usually in schools, over both high- and low-relief rocks. **Description**: Whitespeckled rockfish are small, elongate fish covered with fine white speckles. The body color ranges from red to quite pale and those fish lying on the bottom often have reddish or orangish vertical bars and many fish have a dark red stripe along the flanks. The lateral line is in a clear area that is usually red, but may also be pale.

SIMILAR SPECIES
Starry rockfish have many white spots, but also have 5–6 large white blotches on the back and are deeper bodied. Dwarf-red rockfish lack the white speckles and are not as slim.

Juvenile.

SWFSC ROV TEAM

SWFSC ROV TEAM

SWFSC ROV TEAM

99

Sebastes rufinanus Lea & Fitch, 1972
Dwarf–red Rockfish

Maximum Length: 17.2 cm (6.8 in) TL. **Geographic Range**: San Miguel Island (33°59'N, 120°22'W), southern California to 60 Mile Bank (32°00'N, 118°12'W). **Depth Range**: 58–220 m (191–722 ft). **Habitat**: This diminutive species generally lives very close to the bottom and often forms small aggregations. It lives over a variety of sea floors, although we see the majority of them over high relief and cobble. **Description**: Dwarf-red rockfish are small, slim fish, with reduced spines. Confusingly, despite their name, when underwater dwarf–red rockfish are not always red. Rather they are most often a drab pink-tan, pink-gray, or dusky-red, with a somewhat darker back and slightly lighter sides and belly. A distinguishing character is the prominent whitish, pink, or reddish lateral line. Linda Snook tells us that the white-edged lower pelvic fin rays are also usually characteristic.

SIMILAR SPECIES
Whitespeckled rockfish are red and profusely covered in small white dots and a broad red lateral line. Juvenile squarespot rockfish are tan, more oval, and the lateral line is not prominent.

The fish on top is a whitespeckled rockfish.

Sebastes emphaeus (Starks, 1911)
Puget Sound Rockfish

Maximum Length: 19 cm (7.5 in) TL. **Geographic Range**: Outer coast of Kenai Peninsula and Prince William Sound, northern Gulf of Alaska to Point Sur, central California. **Depth Range**: 3–470 m (10–1,542 ft), typically in 10 m (33 ft) and deeper. **Habitat**: This is a benthic and schooling species, found primarily over such high relief as boulders and sheer vertical rock walls. They will come to the surface on occasion. **Description**: Puget Sound rockfish are slim fish with large eyes. We note that the body colors and patterns of this species are quite variable as they can appear tan (washed with yellow and pink), pink-red, brown, and reddish-brown. Adults dwelling in crevices are often a darker orange, and fish at night may be blotchy, while schooling individuals can be very pale. Many fish have a red-brown or brown streak (this is sometimes poorly defined), of various lengths, running under the lateral line. A dark vertical band on the caudal peduncle is often present.

SIMILAR SPECIES
The closely related pygmy rockfish looks somewhat similar to this species. It is similarly shaped, but tends to be various shades of red and pink, and has a white or red streak below the lateral line and a yellow or white belly.

Note dark vertical band and dark streak. ANDY MURCH

PAULINE RIDINGS

JANNA NICHOLS

Night coloration.

JANNA NICHOLS

MARC CHAMBERLAIN

103

Sebastes wilsoni (Gilbert, 1915)
Pygmy Rockfish

Maximum Length: 24 cm (9.4 in) TL. **Geographic Range**: Kodiak Island, western Gulf of Alaska to 60 Mile Bank (32°01'N, 118°13'W), southern California; they are abundant as far southwards as southern California. **Depth Range**: 30–270 m (98–891 ft), although most fish live in 60–150 m (197–492 ft). **Habitat**: The adults of these schooling fish are usually found over structure and often over high relief. The juveniles often school over sand or mud near rocks. **Description**: Pygmy rockfish are small and semi-elongated with a distinct diamond shape. *Juveniles*: Young-of-the-year are buff or tan with slightly darker mottling on back and sides and the dorsal fin is heavily spotted. However, from a distance these young and quite small individuals appear to have no markings. The characteristic broad and semi-diffuse red or brown flank stripe develops when these fish are about 6 cm (2.4 in) long. *Adults*: Adults range in color from tan or pink to orange or red; the more vibrant individuals have a yellow lower side and belly. Similar to a number of other rockfish species, pygmies lying on the sea floor develop extensive blotching and saddling. There appears to be at least two "forms" of this species. One, a "white-stripe" form is a pinkish-red fish with a white stripe along the lower sides and a white belly. The other type ("red-stripe") is variable in color, but tends to be blotchy red, often with a distinct red stripe on the lower sides and a yellow or white belly. There are images of "white stripes" from Cordell Bank, northern California, and off Washington State and southeastern Alaska. "Red-stripes" live at least in southern and central California. Given the propensity of rockfishes to speciate, we wonder if these "forms" are actually distinct species.

Young-of-the-year.

Young-of-the-year.

SIMILAR SPECIES

The red or white stripe on the lower flanks of pygmies and their distinct elongated diamond shape differentiate it from virtually all other northeastern Pacific rockfishes, with the possible exception of the closely related Puget Sound rockfish. Puget Sounds are similarly shaped, but are brown or orange-brown fish often with a dark brown stripe on the flanks and a dusky vertical bar on the caudal penduncle (just anterior of the tail fin).

Red-stripe form.

Red-stripe form.

White-stripe form.

Red-stripe form.

Sebastes elongatus Ayres, 1859
Greenstriped Rockfish

Maximum Length: About 48.7 cm (19.2 in) TL. **Geographic Range**: Chirikof Island, Alaska (55°50'N, 155°37'W) to Isla Cedros (28°13'N, 115°15'W) and Isla Guadalupe, central Baja California, abundantly from Yakutat, Gulf of Alaska to at least Bahia San Quintin, northern Baja California. **Depth Range**: 12–1,145 m (39–3,756 ft) and mostly in 100–300 m (328–984 ft). **Habitat**: Although the elongated body would suggest a fish that occupies the water column, greenstriped rockfish are usually found resting on the bottom. Young fish recruit from the plankton to such low and hard relief as cobble or the shell mounds around oil platforms and also to mud. Both juveniles and adults are usually solitary and live primarily over mixed mud and low-relief rocks and shell mounds, but not uncommonly on both mud sea floors and boulders. **Description**: The greenstriped rockfish is a relatively streamlined species that is white, pink, or red and has four large irregular green stripes on the back and sides. Andy Lamb describes these as "rows of a series of closely aligned blotches." The lateral line is in a clear area and there are two stripes above and below that line. The lips are brushed with red.

Juvenile.

SIMILAR SPECIES

With their series of dark green stripes, greenstripeds are fairly iconic; it is unlikely that they will be confused with any other northeastern Pacific species. Other green-marked rockfishes, such as greenspotteds and greenblotcheds have a series of distinct lighter patches on the back and are much more robust. In addition, greenspotteds have green spots on the back, while greenblotched carry irregular green vermiculations.

SWFSC ROV TEAM

SWFSC ROV TEAM

SWFSC ROV TEAM

SWFSC ROV TEAM

Sebastes babcocki (Thompson, 1915)
Redbanded Rockfish

Maximum Length: 71 cm (28 in) TL. **Geographic Range**: Bering Sea at Zhemchug Canyon and Amchitka Island, Aleutian Islands to San Diego, southern California, most abundant from about the Kodiak Island area eastward and southward to central California. **Depth Range**: 31–1,145 m (102–3,756 ft); primarily in 150–450 m (492–1,476 ft). **Habitat**: Redbandeds are found both near and on the sea floor around pinnacles and high-relief rock and also over flat rocky areas. We usually see them as solitary individuals. **Description**: These are fairly deep-bodied fish, with strong head spines, a rounded snout, and four dark-red, orange-red, or (in young-of-the-year) reddish-brown vertical bands on the body. The soft dorsal and anal fins are often black-streaked and white-edged. A key character is the first vertical band that runs from the anterior part of the spiny dorsal fin diagonally backwards, just grazing the rear of the gill cover and continuing behind the pectoral fin. Late-stage pelagic juveniles, similar to splitnose at the same life stage, lack obvious bands.

SIMILAR SPECIES
Along with redbandeds, three other northeastern Pacific rockfish species have vertical bandings (e.g., flag and tiger

Young-of-the-year.

rockfishes and treefish). Flag rockfish look very similar to redbandeds and for many decades the two were considered to be one species. However, flags have a more pointed snout and, perhaps of most importance, their first red bar is vertical and runs down over the gill cover, rather than extending backwards, barely over the rear of the gill cover, as in redbandeds. Tiger rockfish have five vertical bands that are thinner than those of redbandeds. These are dark red, purple, brown, or black, overlaying a pink or white body. Treefish are banded (usually seven dark bands compared to four) black, brown, or dark green alternating with yellow or light green; they have pink or red lips. Juvenile cowcod also have bars, but these are irregular or broken.

Note position of bar.

Sebastes rubrivinctus (Jordan & Gilbert, 1880)
Flag Rockfish

Maximum Length: 44 cm (17.2 in) TL. **Geographic Range**: Heceta Bank, Oregon (44°05'N, 124°50'W) to southern Bahia de Sebastian Vizcaino, central Baja California (28°06'N), typically from central California to at least off Bahia San Quintin, northern Baja California. **Depth Range**: 30–431 m (100–1,414 ft), in southern California mostly in 60–160 m (197–525 ft). **Habitat**: Flags are benthic and often solitary fish (although we have, on occasion, seen them in small aggregations). Juveniles recruit to drifting kelp mats, oil platforms, and other hard substrate. Adults mostly inhabit areas with caves and crevices, smaller ones can be found over cobble. Occasionally an individual will be found over soft sea floors adjacent to hard structure. **Description**: Flag rockfish are somewhat ovate and spiny fish, characterized by alternating white and four, wide, orange, red or (in young-of-the-year) reddish-brown bars, and a fairly pointed snout. The most forward bar passes in front of the pectoral fin and over the rear of the gill cover. A stripe extends from the top of the head through the eye and then splits, one branch running to the rear of the lower jaw and the other anterior along the snout. The soft dorsal, caudal, and anal fins are rimmed in white.

Young-of-the-year.

Juvenile.

SIMILAR SPECIES

There are three other northeastern Pacific rockfishes with intense vertical barring: redbanded, tiger, and treefish. Of these, redbanded rockfish can be most easily confused with flags as they have very similar coloration (and, for many years, were thought to be the same species). For redbandeds, two key differences, visible both underwater and after capture, are 1) the first vertical red bar runs from the anterior part of the spiny dorsal fin diagonally backwards and just grazes the rear of the gill cover (it covers much more of the gill cover in flags) and passes behind the pectoral fin base, and 2) redbandeds have a more rounded snout. Tiger rockfish also have vertical bars; these are often black, brown, or purple, but red-barred tigers do exist. However, tigers have five (rather than four) bars on the body, and these are thinner than those on flags. Treefish have red or pink lips and usually have seven dark bars; these are black, brown, or dark green alternating with green or yellow-green. Juvenile cowcod are also barred, but the bars are irregular and broken.

Note position of forward bar; it passes in front of the pectoral fin and over the rear of the gill cover.

Sebastes nigrocinctus Ayres, 1859
Tiger Rockfish

Maximum Length: 61 cm (24 in) TL. **Geographic Range**: Off Eider Point, Unalaska Island, Aleutian Islands to Tanner and Cortes banks, southern California, typically from southeastern Alaska to northern California. **Depth Range**: 9–298 m (30–978 ft), adults commonly are in 30 m (98 ft) and deeper. **Habitat**: Little is known about the habitats of juveniles, although some young-of-the-year have been found around such floating objects as floats and drift kelp (where their colors match the surrounding algae). Adults are often solitary and territorial, mostly reclusive, and live in crevice-filled high-relief rocks. **Description**: Tiger rockfish are handsome, deep-bodied, spiny, and craggy fish, characterized by black, brown, red, or purple vertical bars overlaying a white or pink body. Juveniles look similar to adults, although they tend to have small spots in between the bars.

SIMILAR SPECIES
Treefish usually have black, brown, or dark green bars alternating with green or yellow-green; they have red or pink lips. Flag and redbanded rockfishes also have vertical bars. However, they have four bars on the body and these are wider than those on tigers and always red or red-orange. Juvenile cowcod also have bars, but these are irregular and broken.

Young-of-the-year.

Juvenile.

Juvenile.

SWFSC ROV TEAM

MARC CHAMBERLAIN   JANNA NICHOLS

MARCOS GUIMARAES

Sebastes serriceps (Jordan & Gilbert, 1880)
Treefish

Maximum Length: 41 cm (16 in) TL. **Geographic Range**: San Francisco, northern California (37°50'N, 121°56'W) to Isla Cedros, central Baja California primarily from southern California to northern Baja California. **Depth Range**: Intertidal–106 m (0–348 ft), mostly in 50 m (165 ft) and less. **Habitat**: Benthic and solitary (and territorial), treefish occupy a variety of complex relief habitats, including kelp forests, rocky reefs, and oil platforms. **Description**: Like many other benthic and sedentary species, treefish are spiny and squat (with particularly prominent head spines). This species has a pointed snout and, as an adult, six vertical bars on the body, the first bar, located below the anterior part of the spiny dorsal fin, is partially split in two. *Juveniles*: Younger fish have double bars, these fill in as the fish mature. Juveniles have an additional vertical bar on the caudal fin, along with dark anal, soft dorsal, pectoral, and pelvic fins. Young-of-the-year living around drifting material, such as kelp mats, are light yellow dorsally, becoming silvery below. Benthic juveniles tend to be bright yellow with black bars. *Adults*: These are yellow or green-yellow, and have black, brown, or dark-green bars, and pink or red lips. Many adults are covered in very small light speckling.

Juvenile.

SIMILAR SPECIES
There are three other northeastern Pacific rockfishes with intense vertical bars; tigers, flags, and redbandeds. Of these, only tiger rockfish might easily be mistaken for treefish. Tigers have similar dark vertical barring, but the species background color is pink, tan, or orange. Flag and redbanded rockfishes also have vertical bars. However, they have four bars on the body and these are always red, red-orange or (in young juveniles) red-brown. Juvenile cowcod also have bars, but these are irregular or broken.

Juvenile. An unusual color and pattern variant.

SWFSC ROV TEAM

DAVID ANDREW

SWFSC ROV TEAM

SWFSC ROV TEAM

Sebastes maliger (Jordan & Gilbert, 1880)
Quillback Rockfish

Maximum Length: 63.6 cm (25 in) TL. **Geographic Range**: Kodiak Island, Gulf of Alaska to Anacapa Passage, southern California, most abundantly from southeastern Alaska to central California. **Depth Range**: Intertidal to 274 m (0–899 ft), typically from perhaps 10–130 m (33–426 ft). **Habitat**: Quillbacks are generally benthic-oriented fish that live in complex habitats. Juveniles tend to live around such habitats as rocks, eelgrass, kelp, and cloud sponge gardens. Adults live mostly over high relief and next to vertical walls and are often solitary, but will sometimes aggregate in moderate-sized groups. For instance, our associates Lynne Yamanaka and Rick Stanley saw an aggregation of perhaps 100 fish off the Queen Charlotte Islands of British Columbia. **Description**: Quillback rockfish are squat, thick-bodied fish that are distinguished by a very high spiny dorsal fin with deeply incised membranes. They are brown or orange-brown, with white, yellow, or orange lighter areas on the head and on the back below the spiny dorsal fin.

Juvenile. MARC CHAMBERLAIN

Qullbacks become progressively darker towards their posteriors. Most of the dorsal spines are also light colored. Brown or orange flecking dots the head and anterior portion of the body of many, although not all, individuals. Juveniles are profusely covered in brown to red spots. **Note:** Declared "threatened" by the Committee on the Status of Endangered Wildlife in Canada.

Juvenile. PAULINE RIDINGS

SIMILAR SPECIES

A few of the benthic, solitary, and spiny rockfishes may be confused with this species, although none of them have the deeply incised margins in the spiny dorsal fin. China rockfish are easily distinguished by color (black or blue-black and yellow) and the prominent yellow band that runs down from the spiny dorsal fin and along the lateral line to the caudal fin. Copper rockfish have a clear zone surrounding the lateral line and gopher rockfish are heavily marked with patches of black, brown, or olive, alternating with pink, flesh, or white.

JANNA NICHOLS

MARC CHAMBERLAIN

KEITH CLEMENTS

Sebastes nebulosus Ayres, 1854
China Rockfish

Maximum Length: 45 cm (18 in) TL. **Geographic Range:** Kodiak Island, western Gulf of Alaska to Redondo Beach and San Nicolas Island, southern California, typically from at least as far north as British Columbia to central California. **Depth Range:** 3–128 m (10–420 ft), most fish live in 10 m (33 ft) and greater. **Habitat:** Chinas are benthic and solitary (and territorial) fish that live in such complex structure as boulder fields or kelp beds. **Description:** A lovely and uniquely patterned squat and spiny species. *Juveniles*: Newly settled young-of-the-year have black or blue-black and bright yellow blotches with white splotches throughout. The snout is bright yellow. *Older Juveniles and Adults:* These are black or blue-black with a prominent yellow stripe that runs down from near the front of the spiny dorsal fin along the lateral line to the base of the tail. The head is dark with yellow patches and some individuals are profusely marked with small yellow or white spots.

SIMILAR SPECIES

Young-of-the-year black-and-yellow rockfish can be confused with similar-sized Chinas, and we are not completely sure we can always separate the species when they are very small. However, it is our sense that very young recruited Chinas have yellow snouts (as opposed to darker one in black-and-yellows), black or blue-black and bright yellow blotching (mostly brownish and drabber yellow in black-and-yellows), and they soon harbor a yellow lateral stripe and/or a myriad of spots (both absent in black-and-yellows). Also, the yellow 'Y' pigment originating at the top of the nape is thick in Chinas and thin in black-and-yellows. Chinas have white splotches in the dark patches and black-and-yellows have none.

CLINTON BAUDER   CLINTON BAUDER

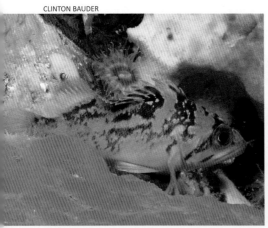

Juvenile China.   Juvenile black-and-yellow.

Adult black-and-yellows can be similar in color, but lack the thick line of yellow pigment extending from spiny dorsal fin to tail. Quillback rockfish have a similar morphology; however, their spiny dorsal fin is much more prominent and the membrane between the spines is much more deeply incised. They tend to be brown or orange-brown and lack both the distinctive yellow band and the many small yellow or white spots.

KEITH CLEMENTS

CSUMB MARE

MARC CHAMBERLAIN

KEITH CLEMENTS

Sebastes chrysomelas (Jordan & Gilbert, 1881)
Black–and–Yellow Rockfish

Maximum Length: 38.7 cm (15.3 in). **Geographic Range**: Cape Blanco (42°50'N, 124°35'W), Oregon to Bahia Santa Maria, (24°46'N, 112°15'W), southern Baja California and common from about Fort Bragg, northern California, to at least northern Baja California. **Depth Range**: Intertidal to 37 m (0–120 ft), typically from 2–15 m (6–50 ft). **Habitat**: Black-and-yellows are benthic and territorial. They live over complex habitat, including both high- and low-relief rocks and in kelp forests. **Description**: Black-and-yellows are squat, spiny, and heavy-bodied. *Juveniles*: Newly recruited benthic juveniles, less than about 7–10 cm (3–4 in), have gold or brown vertical bars over a clear, white, or pale gold background. *Older Juveniles and Adults*: These fish have three or more light patches on the back (extending on to the dorsal fin) and irregular lighter markings along the sides. The darker portions on the body are black or otherwise dark and the lighter ones are yellowish. The posterior part of the lateral line is usually yellow. The anterior part of the lower jaw is also yellow.

SIMILAR SPECIES

We have found that, underwater, juvenile black-and-yellows, less than 7–10 cm (3–4 in) long, cannot, with certainty, be differentiated from gopher, copper, and kelp rockfishes. At about this length, the vertical barring disappears, and black-and-yellows develop dark and light patches, allowing differentiation from coppers and kelps. Young-of-the-year black-and-yellow rockfish can also be confused with similar-sized Chinas, and we are not completely sure we can always separate the species when they are very small. However, it is our sense that very young recruited Chinas have yellow snouts (as opposed to darker ones in black-and-yellows), a bright yellow base coloration, black or blue-black and bright yellow blotching (mostly brownish and drabber yellow in black-and-yellows), and Chinas soon harbor a yellow lateral stripe and/or a myriad of spots (both absent in black-and-yellows). It is not until perhaps 15 cm (6 in) that a black-and-yellow's coloration differs substantially from that of gophers. Gopher rockfish are very closely related and, with the exception of some subtle genetic differences, juveniles and adults are apparently identical in every way except for color. Gophers have the same pattern of mottling, but the light areas are pink, flesh, or white and the darker mottling tend to be brown. China rockfish are similar in coloration, but have a very distinctive wide yellow swath that runs from near the front of the spiny dorsal fin, along the lateral line, to the base of caudal fin.

Juvenile.

Juvenile.

Sebastes carnatus (Jordan & Gilbert, 1880)
Gopher Rockfish

Maximum Length: 39.6 cm (15.6 in) TL. **Geographic Range**: Cape Blanco, Oregon (42°50'N, 124°35'W) to Punta San Roque (27°12'N, 114°26'W) and typically from northern California to at least off Bahia San Quintin, northern Baja California. **Depth Range**: Intertidal to 86 m (0–282 ft) and mostly in about 12–50 m (39–164 ft). **Habitat**: A substrate-associating solitary (and territorial) species, both juveniles (who recruit to kelp canopies and other substrata) and adults are found primarily over high relief in such habitats as rocky reefs and kelp forests. **Description**: This is a squat, spiny, and heavy-bodied species. *Juveniles*: Newly recruited benthic juveniles, less than about 7–10 cm (3–4 in), have five gold or brown vertical bars over a clear, white, or pale-gold background. These bars gradually break up into the familiar patches of older fish. *Older Juveniles and Adults*: These have three or more light patches on the back (extending on to the dorsal fin) and irregular lighter areas along the sides. The darker parts are black, brown, or olive and the lighter ones are pink, flesh, or white, and the anterior of the lower jaw is often yellowish. One dark stripe radiates back from the eye and another runs the same way from the upper jaw.

SIMILAR SPECIES

Underwater, juvenile gopher rockfish less than 7–10 cm (3–4 in) long cannot, with certainty, be differentiated from black-and-yellow, copper, and kelp rockfishes. At about this length, gophers lose their vertical barring and develop their familiar patches, allowing differentiation from coppers and kelps. However, it is not until perhaps 15 cm (6 in) that a gopher's coloration varies from that of black-and-yellows. Turning to older juveniles and adults, black-and-yellow rockfish (which are extremely closely related to gophers) are virtually identical

Young-of-the-year.

in appearance to gophers, with similar body morphology and body markings. The only apparent differences, other than subtle genetic ones, are those of color. The darker markings on black-and-yellows are black or olivaceous, while the lighter portions are yellow. With their somewhat similar marking pattern, copper rockfish are another species that are frequently mistaken for gophers. However, coppers have a sharper snout, a characteristic wide clear band along the lateral line from mid body to the caudal fin (gophers may have a narrow clear band along the rear part of the flanks), and their dark backs tend to be less broken up with blotches. In addition, while their colors are highly variable, we have never seen the pink color in coppers that some gophers assume. Quillbacks have a higher and much more deeply incised spiny dorsal fin and lack the light and dark patches. They become progressively darker toward the posterior and are often very dark brown or black on the tail and caudal peduncle. Calico rockfish have a series of distinct reddish-brown bands that slant back from the base of the dorsal fin through the lateral line almost to the belly.

A fairly unusual, kind of stippled, individual.

Sebastes caurinus Richardson, 1844
Copper Rockfish

Maximum Length: 66 cm (26 in) TL. **Geographic Range**: Western Gulf of Alaska east of Kodiak Island to Islas San Benito (28°18'N, 115°33'W), and mostly from southeastern Alaska to at least off Bahia San Quintin, northern Baja California. **Depth Range**: Intertidal zone to 408 m (0–1,338 ft), characteristically from the shallow subtidal (particularly in the northern part of the range) to perhaps 70 m (230 ft). **Habitat**: As both juveniles and adults, coppers are benthic and usually complex-habitat favoring fish (although they will live over soft sediment on occasion). Although they are often observed as solitary individuals, we have seen them in loose aggregations, particularly around oil platforms.

Young-of-the-year.

Description: Copper rockfish are deep bodied with a relatively high spiny dorsal fin and a relatively sharp snout. *Juveniles*: Young-of-the-year benthic fish, less than about 7–10 cm (3–4 in), have five gold or brown vertical bars over a clear, white, or pale gold background. Within a few months of settling, coppers develop a clear area over the posterior two-thirds of the lateral line. *Older Juveniles and Adults*: The coloration of older juveniles and adults is highly variable, ranging from almost white, yellowish–brown or red to dark brown or almost black. In particular, we have noted that fish in the more northerly part of the range are often darker than those to the south. There is a swath of darker color on the back above the lateral line, lightly broken up with lighter patches. Underwater, all of the coppers we have observed have a clear area along the lateral line from the middle of the spiny dorsal fin to the caudal fin; this clear area may not be visible in dead specimens. One stripe runs diagonally back from the eye and one from the region of the upper jaw. We note that, in this species, individual scales tend to stand out compared to those of many other species.

Young-of-the-year.

SIMILAR SPECIES

When newly settled, coppers cannot be differentiated, underwater, from kelp, gopher, and black-and-yellow rockfishes. However, by about 7–10 cm long, coppers have lost their vertical barring and developed a light-colored lateral line that distinguishes them from other species. We find that many divers confuse older juvenile and adult coppers with gopher rockfish and with good reason, as the two species share a similar shape and markings. There are, however, several reasonably good differentiating characters: 1) Gophers tend to be more heavily blotched on back and sides. 2) Although the rear part of the gopher rockfish lateral line tends to be pink or white, it is, nonetheless, broken up with darker pigment. Lateral lines of copper rockfish are generally completely clear. 3) Lastly, while there is often an almost uninterrupted swath of color on the backs of copper rockfish, those of gophers have a number of lighter blotches. Quillback rockfish may also appear similar to coppers. Quillback spiny dorsal fins are much more deeply incised and there is no clear zone around the lateral line. Calico rockfish have a number of red-brown or brown bars on the back or dorsal fins that continue down the sides.

BETTY BASTAI

SWFSC ROV TEAM

SWFSC ROV TEAM

SWFSC ROV TEAM

SWFSC ROV TEAM

Sebastes auriculatus Girard, 1854
Brown Rockfish

Maximum Length: 56 cm (22.4 in) TL. **Geographic Range**: Prince William Sound, Alaska (59°32'N, 151°36'W) to Bahia Magdalena, southern Baja California. Primarily from southeastern Alaska to Puget Sound and from Bodega Bay, northern California to Bahia Tortugas, southern Baja California. **Depth Range**: Surface to 294 m (0–964 ft), typically in perhaps 70 m (230 ft) and less. **Habitat**: We usually see these fish as solitary individuals (occasionally in small aggregations), usually near or on the bottom, over low and high relief rocks, oil platforms, and mud walls of submarine canyons. They also inhabit murky bays. **Description**: Brown rockfish are medium-heavy bodied, and spiny headed. They are tan or reddish brown, with numerous dark markings, a dark mark on the back upper corner of the gill cover, and the eye often has a red or orange cast. Two orange or orange-brown lines radiate backwards from the region of the upper jaw and eye.

SIMILAR SPECIES

Grass rockfish are similarly shaped and can be confused with browns. However, grasses tend to be green or greenish-brown and lack the obvious dark spot on the rear of the gill cover. Kelp rockfish may be the same color, but they are slimmer, have larger and lacier pectoral fins, and lack the distinctive dark spot on the gill cover (however, they may have a faint one on the gill cover flap). Kelp rockfish have 31–34 gill rakers while brown rockfish have only 25–29. Calico rockfish have a series of red-brown or brown bars that originate on the back and dorsal fins and continue down the sides.

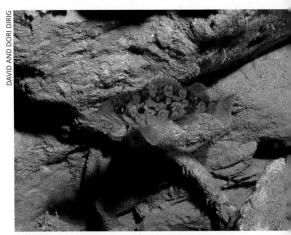

Juvenile.

Note dark blotch.

Nice stogie, brownie.

Sebastes rastrelliger (Jordan & Gilbert, 1880)
Grass Rockfish

Maximum Length: 55.9 cm (22 in) TL. **Geographic Range**: Ucluelet, Vancouver Island, British Columbia to Bahia Playa Maria (28°50'N), central Baja California typically from perhaps northern California to as far southward as central Baja California. **Depth Range**: Intertidal area and to 46 m (0–150 ft), commonly from the intertidal to 15 m (50 ft). **Habitat**: These are usually solitary fish, always found in such complex habitat as tide pools, kelp beds, rocky reefs, and oil platforms. **Description**: Grasses are thick-bodied and spiny, with blunt and rounded snouts, clearly the Bubba of nearshore rockfishes. They are green to brown-green, mottled and splotched with darker pigmentation, and there is often a wide dark line running backwards from the posterior part of the eye. The eyes are relatively small and tend to have a thin yellowish rim. Juveniles are similarly colored and marked.

Young-of-the-year.   STUART HALEWOOD

SIMILAR SPECIES

With their similar body shape and dark splotches, brown rockfish are the species most likely to be mistaken for grasses. However, browns tend to be brown, rather than green, have orangish lines running backwards from upper jaw and eye, and have a distinct dark blotch on the rear of the gill cover (grasses may have a diffuse dark area in the same region). Their eyes often have a reddish outer rim. Kelp rockfish might also confuse the observer. They are thinner bodied, have more pointed snouts and lacier pectoral fins, less distinctive dark spots, and larger eyes.

SCOTT GIETLER

BOB WOHLERS

CHRIS GROSSMAN

DAN RICHARDS

Sebastes dallii (Eigenmann & Beeson, 1894)
Calico Rockfish

Maximum Length: 25.4 cm (10 in) TL. **Geographic Range**: San Francisco, northern California to Punta Rompiente, southern Baja California commonly from at least Santa Cruz, central California through Bahia de Sebastian Vizcaino, central Baja California. **Depth Range**: Intertidal to 305 m (0–1,000 ft), and typically at about 45–70 m (148–230 ft). **Habitat**: This is a mostly benthic species, found either singly or in small, diffuse aggregations. They primarily live on habitat composed of a mixture of low, hard relief and on the surrounding sand and mud. We see large numbers of them on some of the shell mounds surrounding oil platforms. **Description**: Calico rockfish are compact and diminutive fish. They are brown or yellowish-green and generously flecked with red-brown and brown and the caudal fin is striated with brown or red-brown. However, they are best characterized by a series of red-brown or brown bars that originate on the back (and flow up into the dorsal fins) and continue either vertically, or slant obliquely, down the sides.

Juvenile.

SIMILAR SPECIES

Based on the queries that we receive, calico and brown rockfishes are often confused with each other. This is particularly true of juvenile browns, which can be heavily mottled with darker brown and have a somewhat similar shape, thus bearing a passing resemblance to calicos. However, browns do not have the well-organized red-brown or brown barring and do have an isolated dark mark on the gill cover that is absent in the calico rockfish (the first bar on a calico's head may reach onto the gill cover, appearing at first glance to be a blotch). Gopher rockfish have discrete tan and pink blotches on their sides and copper rockfish are also often blotchy and have a clear lateral line.

KEVIN LEE

SWFSC ROV TEAM

Sebastes atrovirens (Jordan & Gilbert, 1880)
Kelp Rockfish

Maximum Length: 42.5 cm (16.8 in) TL. **Geographic Range**: Fort Bragg, northern California (39°27'N, 123°48'W) to Bahia San Carlos (29°36'N, 115°12'W) and Islas San Benito, central Baja California and abundant from around Point Reyes, northern California and southward. **Depth Range**: 0–82 m (269 ft), typically in less than 30 m (98 ft). **Habitat**: Kelp rockfish live in complex habitats including kelp forests and over both low and high relief rocks, both on the bottom and associated with midwater substrata (such as giant kelp plants). We see them both as solitary individuals and in aggregations. **Description**: Kelp rockfish are thin, fairly oval fish with a relatively sharp snout, large eyes, and lacy, and generally translucent, large pectoral fins. *Juveniles*: Newly recruited benthic juveniles [young-of-the-year], less than about 7–10 cm (3–4 in), have five brown vertical bars over a clear, white, or pale brown background. Within a short time, these young fish develop a very characteristic "Y" shape to the vertical bar under the soft dorsal fin. As these fish develop, the barring becomes much more diffuse and is replaced by speckling which may form faint vertical columns. *Older Juveniles and Adults*: These can range in color from almost white through various shades of brown, yellow, and even reddish. Much of the body is covered in darker flecking and there may be a faint dark area on the posterior flap of the gill cover.

SIMILAR SPECIES

We know of no way to distinguish, underwater, very early, newly settled, young-of-the-year kelp rockfish from young, and similar sized, coppers, gophers, and black-and-yellows. However, within a short time, the "Y" pattern emerges on the fourth vertical bar and by about 7–10 cm (3–4 in) long, kelps have developed their typical body shape and patterning. Older kelps are most often mistaken for grass rockfish. Grass rockfish are thicker in the head and heavier bodied, usually greenish, are covered in small black spots (these are more distinctive than in kelps), and have short stubby gill rakers. Brown rockfish also kind of resemble kelps; however, they are also thicker and deeper bodied, are brown to red-brown, have orangish lines radiating backwards from the eyes, and always carry a large dark spot on their gill covers. The pectoral fins of both grasses and browns lack the laciness of kelp rockfish.

Young-of-the-year. Note "Y" pattern.

Juvenile.

Note large eye and faint dark area on gill cover.

This image nicely illustrates the thin body and lacy pectoral fins.

Sebastes baramenuke (Wakiya, 1917)
Brickred Rockfish

Also in Asia

Maximum Length: 51.7 cm (20.3 in) TL. **Geographic Range**: Sea of Japan off Korean Peninsula and Pacific Ocean off Honshu, Japan to Kiska Pass, western Aleutian Islands (51°57'N, 176°32'E), Alaska. **Depth Range**: 100–760 m (328–2,493 ft). **Habitat**: There are no published details about this species' behavior and habitat. **Description**: This is a deep-bodied and spiny species. After capture (we have not seen this species underwater), they are red with a dark splotch on the gill cover, two darker stripes running diagonally backwards from the eye, and often with darker (sometime greenish) bars on the head and back. The most diagnostic character is the three dark bands that run across the top of the head. These are placed 1) between the anterior parts of the eyes, 2) at the posterior of these eyes, and 3) just anterior of the dorsal fin.

SIMILAR SPECIES
Although a number of species, such as aurora, chameleon, and splitnose may bear some resemblance to the brickred rockfish, they all lack the three dorsal head bands of this species.

Sebastes glaucus Hilgendorf, 1880
Gray Rockfish

Also in Asia

Maximum Length: 59 cm (23.2 in) TL. **Geographic Range**: Northern Sea of Japan and Sea of Okhotsk to Commander Islands and western Bering Sea north to Amayan Bay, Koryak coast (about 60°50'N); south of Atka Island, Aleutian Islands and Bering Sea. Only a few individuals have been taken off Alaska. **Depth Range**: 2–550 m (7–1,804 ft). **Habitat**: The little that is known about this species implies that it is found over both high-relief rocky bottoms and over soft sea floors. **Description**: We have not observed this species in life. Based on the images we have examined and published descriptions, this is a modestly oval fish with reduced head spines and 14 dorsal spines. It appears to be gray, brown, or black, with body and fins heavily washed and mottled in yellow or orange-yellow. At least three dark bars (two forming a "<") radiate from the eyes.

SIMILAR SPECIES

We know of no similar species in the northeastern Pacific. The northern rockfish also has 14 dorsal spines, however it is quite slim and very differently colored and marked.

Thornyheads

SWFSC ROV TEAM
SHORTSPINE THORNYHEAD

Sebastolobus alascanus Bean, 1890
Shortspine Thornyhead

Also in Asia

Maximum Length: 82.6 cm (32.5 in) FL. **Geographic Range**: Seas of Okhotsk and Japan to Pacific Ocean and Bering Sea off Kamchatka to Navarin Canyon and Aleutian Islands to Boca de Santo Domingo (25°32'N, 113°04'W), southern Baja California; typically from the northern Kuril Islands, through the Aleutians, in some areas of the Bering Sea, and southward at least as far as southern California. **Depth Range**: 17–1,524 m (56–5,000 ft); most fish are found in about 200 to possibly 800 m (656–2,624 ft). Fish in the northern part of their range tend to live in shallower waters than do those to the south. **Habitat**: Juveniles live primarily over mud bottoms, sometimes near small rocks and other solid objects. We see adults on mud (often near rocks or debris), and over low-hard relief, or even, rarely, over boulder fields. They are usually solitary and almost always lying right on the sea floor. **Description**: Shortspine thornyheads are large-headed, but slim bodied, fish characterized by a row of 8–10 strong spines along the cheek. Both juveniles and adults are red or pink, with greater or lesser amounts of white speckling or blotching. A key character is that the third dorsal spine is not elongated, as it is in the longspine thornyhead, rather the fourth or fifth spine is slightly longer than the others. The pectoral fins of newly settled juveniles through small adults are often speckled with white.

Juvenile.

SIMILAR SPECIES
The depth range and habitats of shortspines and longspine thornyheads greatly overlap and, underwater, the two species may look very similar. Newly settled juvenile longspines can be differentiated by their black pectoral fins. Unlike in shortspines, the pectoral fin of juveniles and small adults are not speckled with white. In addition, unlike in the shortspine, the third dorsal spine of longspines is significantly longer than the fourth or fifth one. In general, separating larger individuals of the two species can be difficult when the spiny dorsal fin is not erect. Broadfin thornyheads have one or two black marks in the posterior part of the dorsal fin and the branched lower pectoral rays are highly branched.

Juvenile.

Juvenile.

141

Sebastolobus altivelis Gilbert, 1896
Longspine Thornyhead

Maximum Length: 38 cm (15 in) TL. **Geographic Range**: Bering Sea (59°20'N, 178°17'W) and Shumagin Island, western Gulf of Alaska to Cabo San Lucas, southern Baja California and Isla Guadalupe, central Baja California, mostly from Vancouver Island to at least southern California. **Depth Range**: 163–1,756 m (535–5,760 ft), typically in 500–1,300 m (1,640–4,264 ft). **Habitat**: These solitary and sedentary fish spend most of their time on mud, often near bits of rocks or other hard material. **Description**: Similar to shortspines, longspine thornyheads are elongated fish with large, spiny heads and with a row of 8–10 spines along each cheek. The most useful character in identifying this species is the clearly elongated third dorsal spine, which is considerably longer than any other. This species is pink, red, or brownish, and the juveniles, in particular, often have white splotches and streaks. Newly settled juveniles have black pectoral fins.

Juvenile.

SIMILAR SPECIES

Shortspine thornyheads look similar to longspines. Newly settled shortspines have reddish pectoral fins, those of longspines are black. The pectoral fins of juvenile and small adult shortspines are white speckled; speckling is absent on longspines. The best character separating adults of the two species are the length of the third dorsal spine – it is not elongated in shortspines. When dorsal spines of the adults are not erect, we have found it difficult to separate these two species. Broadfin thornyheads have one or two black marks in posterior part of the dorsal fin and the lower pectoral rays are highly branched.

Juvenile.

Note elongated third dorsal spine.

Sebastolobus macrochir (Günther, 1877)   Also in Asia
Broadfin Thornyhead

Maximum Length: 38 cm (15 in) TL. **Geographic Range**: Seas of Japan and Okhotsk to Commander Islands, Pacific Ocean south of Aleutian Islands (eastward to 54°22'N, 166°20'W), and Bering Sea south of Cape Navarin to eastern Bering Sea, abundantly in the seas of Japan and Okhotsk to as far east and north as the northern Kurils and southeastern Kamchatka. Only a handful of broadfins have been taken in U.S. waters. **Depth Range**: At depths of 100–1,504 m (328–4,934 ft), typically in 200–800 m (656–2,624 ft). **Habitat:** We have not observed this species. However, based on images from the western Pacific, the habitat and behavior of broadfins is similar to that of both shortspine and longspine thornyheads; broadfins are benthic, solitary fish that live on soft sea floors. **Description**: Broadfins are somewhat slim fish with spiny heads that carry a row of 8–10 strong spines on each cheek. We have not seen these fishes underwater. However, photographs from the western Pacific indicate that they range from brown through reddish to pink. There are one or two black blotches on the posterior of the spiny dorsal fin. The lower rays of the pectoral fin are highly branched.

SIMILAR SPECIES
The fingerlike lower rays of the pectoral fin of this species are unique among the thornyheads. In addition, neither shortspine nor longspine thornyheads have a black blotch on the posterior part of the spiny dorsal fin.

JIM LYLE
CALIFORNIA SCOPIONFISH

Scorpionfishes and ... One Other Fish

Adelosebastes latens Eschmeyer, Abe, & Nakano, 1979
Emperor Rockfish

Maximum Length: 41 cm (13.2 in) TL. **Geographic Range**: North Pacific Ocean in the region of the Emperor Seamounts northward to south of Amilia Island (51°29'N, 173'29'W), Aleutian Islands. Thus far, there are only a handful of records from U.S. waters. **Depth Range**: 352–1,200 m (1,155–3,937 ft). **Habitat**: The habitat of this species is poorly understood. Two of the individuals caught along the Aleutian Islands were taken with such soft sea floor species as sablefish (*Anoplopoma fimbria*), shortspine thornyheads, and skates. **Description**: This is a heavily spined species, characterized by deeply lobed rays on the lower part of the pectoral fins. Generally, emperor's are variations on bright red, often with some dark spotting or blotching; these most often are on the rear part of the gill cover, but also on the back of the head, body, and sometimes fins. Some individuals also have faint brown saddling or banding on the back.

SIMILAR SPECIES
The very deeply lobed, finger-like, rays on the lower parts of the pectoral fin separate the emperor rockfish from almost all other species. In addition, the thornyheads, which do have partially free lower pectoral rays, have 5–10 strong cheek spines (emperors have at most two and they tend to be weak in larger individuals) and 15–18 dorsal spines (there are 12–13 in this species).

Note finger-like rays.

J. W. ORR AND D. E. STEVENSON

We note that there appears to be at least one undescribed *Pontinus* species living along the southern Baja California coast. See Robertson and Allen (2008) for details.

Pontinus furcirhinus (Garman, 1899)
Red Scorpionfish

Maximum Length: 32.4 cm (12.8 in) TL). **Geographic Range**: Isla Guadalupe, central Baja California, and southern Baja California (24°45'N, 115°15'W) to Paita, Peru, including Gulf of California, Islas Galápagos, and Isla Cocos. Frequently caught off southern Baja California. **Depth Range**: 46–390 m (151–1,279 ft). **Habitat**: Benthic, over sand and mud, but also in cobble and boulder fields, occasionally in high-relief rocks. **Description**: Red scorpionfish have compressed, bony, and spiny heads; these lack pits in front of and behind the eyes. The third dorsal spine is elongate; the last dorsal ray is divided down to its base and not attached to the body. Red scorpionfish are red, with some white mottling, and brown or olive spots on upper body. The fins are light and speckled with red or brown spots.

Also at least some offshore islands and Peru

SIMILAR SPECIES
The Peruvian scorpionfish is also red, but does not have the elongated third dorsal spine. The combination of bright red body and, in particular, the long third dorsal spine is sufficient to separate this species from any other scorpionfish.

Pontinus sierra (Gilbert, 1890)
Speckled Scorpionfish

Maximum Length: 28 cm (11 in) TL. **Geographic Range:** Outer coast of southern Baja California to Peru (4°55'S, 81°19'W), including Gulf of California. Apparently rare along the southern Baja California coast. **Depth Range:** About 20–351 m (66–1,151 ft). **Habitat:** High-relief rocky areas. **Description:** This is a relatively slender species. The head is bony and has numerous sharp spines. The eyes are relatively large and the space between the eyes is narrow. There are no particularly elongated dorsal spines and no pits behind the eyes. The body is bright red and there are a series of darker gray or gray-green markings along the back and sides; the fins are spotted with red.

Also to Peru

SIMILAR SPECIES
The combination of a red body with darker dorsal markings, reddish spots on the fins, a lack of both head pits and elongated dorsal spines, separates this species from all others.

JOHN SNOW

ROSS ROBERTSON

Pontinus vaughani Barnhart & Hubbs, 1946
Spotback Scorpionfish

Maximum Length: 57 cm (22.4 in) TL. **Geographic Range**: Isla Cedros, central Baja California to Peru, including southwest part of Gulf of California, Islas Revillagigedo, and Islas Galápagos. **Depth Range**: 30–120 m (98–394 ft). **Habitat**: High-relief rocky areas. **Description**: This is a striking, and quite unique, species. It has a relatively large head and deep body and the second and third dorsal spines are elongated. The head and body are dark-pink or red and the upper parts (as well as the spiny dorsal fin) are washed in yellow. The entire body and all of the fins are densely covered in pale blue spots.

Also at least some offshore islands and Peru

SIMILAR SPECIES
The combination of a myriad of bluish spots and the elongated dorsal spines separates this species from all others.

JOHN SNOW

Scorpaena afuerae Hildebrand, 1946
Peruvian Scorpionfish

Maximum Length: 37 cm (14.6 in) TL. **Geographic Range**: Todos Santos (23°41'N, 110°23'W), southern Baja California, Gulf of California, Costa Rica to Peru, Isla Cocos. Peruvians appear to be relatively common as far north as Todos Santos, southern Baja California. **Depth Range**: 35–100 m (115–328 ft). **Habitat**: High relief. **Description**: The Peruvian scorpionfish has a large, spiny, and depressed head. Importantly, there is a deep depression in front of, and behind, each eye and the ridge below the eye carries 3–4 spines. While there are skin flaps above the front of the upper jaw, there are none on either the lower jaw or body. The body, head, and fins of this species are scarlet to orange-brown with red mottling, there is irregular red barring and darker spotting on fins, and scattered dark spots on the body.

Also Costa Rica to Peru

SIMILAR SPECIES
The red scorpionfish has a somewhat similar color, but has a long third dorsal spine. The California scorpionfish can be a range of colors, but is not scarlet; it is much more heavily marked with black spots and does not have the very deep depressions near the eyes.

JOHN SNOW

Scorpaena guttata Girard, 1854
California Scorpionfish

Maximum Length: 47 cm (18.5 in) TL. **Geographic Range**: Santa Cruz, central California to throughout the Gulf of California, and Isla Guadalupe, central Baja California. Along the Pacific Coast, California scorpionfish are abundant from the Santa Barbara Channel, southern California to Todos Santos, southern Baja California. **Depth Range**: Tide pools to 183 m (600 ft), common from 30–190 m (98–623 ft). **Habitat**: Benthic, on sand, mud, shell debris, low and high-relief rocks and crevices. Although often seen as solitary individuals, during the summer and early fall spawning season, this species forms large aggregations.
Description: The California scorpionfish has a large, spiny, and depressed head containing a number of cirri and a ridge below each eye harboring 0–3 spines. The body color of this species is quite variable, we see them in mixtures of brown, tan, whitish, red, or lavender, and adults always have a myriad of brown or black spots on body and fins. However, note that small juveniles may have few, or, in some instances, no dark spotting.

SIMILAR SPECIES
Rainbow scorpionfish are usually bright red, have a dark blotch on the rear of the gill cover, and have red eyes. The Peruvian scorpionfish has deep depressions behind and in front of the eyes, is bright scarlet, and lacks the myriad dark spots. The player scorpionfish has a dark spot or blotch just above the pectoral fin and lacks the dark spots.

Juvenile.

Juvenile.

Scorpaena histrio Jenyns, 1840
Player Scorpionfish

Maximum Length: 27.3 cm (10.7 in) TL. **Geographic Range**: Isla Guadalupe, central Baja California to Chile, including Gulf of California and Islas Galápagos. Along the mainland outer coast of Baja California, this species is common at least as far north as Todos Santos (23°41'N, 110°23'W), southern Baja California. **Depth Range**: 5–200 m (17–656 ft). **Habitat**: Benthic and solitary, this species lives among hard and high-relief habitats. **Description**: Player scorpionfish have bony, depressed heads with a shallow pit behind each eye. The lateral ridge under each eye has one spine and there are five strong spines on the rear of the gill cover (the top one is largest); the head and body have numerous skin flaps. As with several other scorpionfishes, the body color is variable, but mostly variants of brown and red. Of great importance for identification purposes, there is a dark spot or blotch, ranging from blackish to red or orange, on the flanks above the base of each pectoral fin.

Also to Chile

SIMILAR SPECIES
The California scorpionfish has some superficial resemblance to this species. However, it lacks the distinctive dark mark above the pectoral fin and is liberally covered in dark spots.

Note dark blotch.

JOHN SNOW

Scorpaena mystes Jordan & Starks, 1895
Stone Scorpionfish

Maximum Length: 49 cm (20 in) TL. **Geographic Range**: Redondo Beach, southern California to Chile, including Isla Guadalupe, Gulf of California, Islas Galápagos, and other offshore islands.
Depth Range: Intertidal to 100 m (328 ft). **Habitat**: A solitary and benthic species that is found over a wide range of habitats, from sand and algae to high-relief boulder fields. **Description**: The stone scorpionfish has a very broad and depressed head containing numerous spines; the lateral ridge under each eye houses 3–4 spines. A small pit sits in front of the eye and a deeper one resides behind each eye. Of particular importance in species identification, there are many skin flaps on the head (particularly under the mouth) and on the body. The body color, similar to the case with many other scorpionfishes, is highly variable and the species is capable of rapidly changing both color and pattern. Colors are often drab grays, greens, reds, and browns. A key character is the small white spots that stud the underside of the pectoral fin near its base. In many instances, the caudal fin is barred. Rare, very orange, individuals have a black blotch on the spiny dorsal fin, and do not have distinct caudal fin barring. Young-of-the-year have dark bars and white caudal peduncles.

Also to Chile

SIMILAR SPECIES
The very depressed head, myriad skin flaps under the mouth, and the white spots on the underside of the pectoral fin clearly differentiate this species from all others.

TERRY STRAIT

Note white spots on the underside of the pectoral fin.

Scorpaena russula Jordan & Bollman, 1890
Reddish Scorpionfish

Maximum Length: At least 15 cm (6 in) TL. **Geographic Range**: Bahia Abreojos (26°48'N, 113°25'W), southern Baja California to Chimbote, Peru, including Gulf of California. Apparently relatively rare along the outer coast of Baja California. **Depth Range**: 7–160 m (23–525 ft). **Habitat**: Little is known about this species although it is reputed to live over soft habitats. **Description**: This is a relatively slender species, with a short snout and a narrow and spiny head, that has either no or weak ridges between the eyes. Small cirri are found between the eyes and the lateral ridge below each eye has 2–3 spines. Unlike a number of other scorpionfish species, the body has no skin flaps. While the body color is variable, mottling of brown and red predominate. The pectoral and caudal fins, in particular, are often red, the pectoral fin having a blackish tip. The caudal fin has at least one, and sometimes two, bands of lighter coloration.

Also to Peru

SIMILAR SPECIES
Other relatively slim scorpionfish include rainbow, red, Sonora, and speckled scorpionfishes. Rainbow scorpionfish have a dark spot on the rear of the gill cover, red scorpionfish are bright red and have a long third dorsal spine, a round dark blotch is found on the soft dorsal of Sonoras, and speckleds are bright red, with dark markings on the back and red spots on the fins.

ROSS ROBERTSON

Scorpaena sonorae Jenkins & Evermann, 1889
Sonora Scorpionfish

Maximum Length: 18 cm (7.1 in) TL. **Geographic Range**: Bahia Santa Maria (24°43'N, 112°11'W), southern Baja California and Gulf of California to Guerrero State, Mexico. Apparently, this species is relatively uncommon along the outer coast of Baja California. **Depth Range**: 1–91 m (4–298 ft). **Habitat**: Nothing is known of the habitat or behavior of this species, although it is reputed to live on sand. **Description**: Sonoras are slim fish, with narrow, bony, and spiny heads, short snouts, and 0–2 small spines on the lateral ridge under each eye. The body color is mottled brown, red, and white with dark barring on the sides, there is a large dark blotch on the upper part of the soft dorsal fin, and the caudal fin is vertically banded.

Also to Guerrero State, Mexico

SIMILAR SPECIES
Although there are a number of other slim scorpionfishes in the northeastern Pacific, including rainbow, red, reddish, and speckled, this is the only species with a dark blotch on the soft dorsal fin.

JOHN SNOW

JOHN SNOW

Scorpaenodes xyris (Jordan & Gilbert, 1882)
Rainbow Scorpionfish

Maximum Length: 15 cm (6 in) TL. **Geographic Range**: Anacapa and Santa Barbara islands, southern California to Islas Chincha, Peru, including Isla Guadalupe, central Baja California, Gulf of California, and Islas Galápagos. **Depth Range**: Intertidal to about 50 m (164 ft). **Habitat**: Benthic and solitary, in crevices in high relief. **Description**: Rainbows are relatively slim fish. They have no pits behind the eye, 2–3 spines on the lateral ridge beneath each eye (and sometimes another row of spines below that), and rough scales. There are cirri on the head, but no skin flaps on the body. Rainbow scorpionfish are red and dark brown with white spots, have red eyes, and a dark brown spot on the gill cover.

Also to Peru

SIMILAR SPECIES

The various other slim scorpionfishes, including red, reddish, Sonoran, and speckled, have less rough scales, do not have a prominent dark spot on the gill cover, and lack red eyes. California scorpionfish are occasionally reddish, but are heavier-bodied, much more intensely covered in dark spots, and lack both the dark spot on the gill cover and the red eyes.

ARTURO AYALA BOCOS

KEVIN LEE

Glossary

Anal fin: The unpaired fin located posterior of the anus.

Benthic: Associated with the sea floor.

Caudal fin: The tail fin.

Caudal peduncle: That narrow part of the body to which the caudal fin is attached.

Cirrus (pl., cirri): A hair-like, often branched, structure. In the scorpionfishes, cirri are often located on the head.

Dentigerous knob: A rounded protuberance located on the upper jaw on each side of the snout.

Dorsal fin: The fin(s) located on a fish's back. In the rockfishes, thornyheads, and scorpionfishes, there is one dorsal fin. The anterior is composed of spines and the posterior of rays.

Gill raker: Gill rakers are small, bony processes located on the anterior part of the gill arch. In the rockfishes, thornyheads, and scorpionfishes, gill rakers are used to restrain prey from escaping past the gills.

Lachrymal spine: A spine located just above the upper jaw and slightly in front of the eye.

Lateral line: The line formed by a series of pores located on the body or head.

Lateral ridge: A bony ridge located under the eye.

Meristic: The diagnostic features of a fish, such as fin rays or gill rakers, that can be counted.

Morphometric: The relationship of one body measurement to another (i.e., head length to body length).

Nape: That part of the back of the head located in front of the dorsal fin.

Orbit: The eye.

Pectoral fins: Among the rockfishes, thornyheads, and scorpionfishes, these are paired fins located behind the head on the flanks.

Pelvic fins: The pelvic fins are paired fins located on the underside of a fish's body.

Ray: The segmented, and soft, element of a fin.

Spine: Among the rockfishes, thornyheads, and scorpionfishes, spines are the unsegmented elements of a fin or the sharp projections found on the head and gill cover.

Substratum (pl., substrata): As used in this book, substratum refers to a specific type of habitat on the sea floor.

Symphyseal knob: A rounded protuberance located at the tip of the lower jaw.

Vermiculation: A wave-shaped marking.

Appendix 1
A Comparison of Similarly Appearing Species

Shortbelly rockfish SWFSC ROV TEAM

Chilipepper SWFSC ROV TEAM

Silvergray rockfish SWFSC ROV TEAM

Bocaccio SWFSC ROV TEAM

Mexican rockfish SWFSC ROV TEAM

Bank rockfish SWFSC ROV TEAM

Speckled rockfish SWFSC ROV TEAM

Black rockfish JANNA NICHOLS

Dark rockfish PAULINE RIDINGS

Dusky rockfish SWFSC ROV TEAM

Olive rockfish CLINTON BAUDER

Yellowtail rockfish JANNA NICHOLS

A Comparison of Similarly Appearing Species (continued)

Pacific ocean perch SWFSC ROV TEAM

Sharpchin rockfish SWFSC ROV TEAM

Harlequin rockfish SWFSC ROV TEAM

Redstriped rockfish SWFSC ROV TEAM

Stripetail rockfish SWFSC ROV TEAM

Halfbanded rockfish SWFSC ROV TEAM

Splitnose rockfish SWFSC ROV TEAM

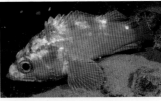
Aurora rockfish SWFSC ROV TEAM

Chameleon rockfish SWFSC ROV TEAM

Canary rockfish SWFSC ROV TEAM

Type 1 Vermilion rockfish SWFSC ROV TEAM

Type 2 Vermilion rockfish SWFSC

A Comparison of Similarly Appearing Species (continued)

Yelloweye rockfish SWFSC ROV TEAM

Cowcod SWFSC ROV TEAM

Bronzespotted rockfish SWFSC ROV TEAM

Greenspotted rockfish SWFSC ROV TEAM

Greenblotched rockfish SWFSC ROV TEAM

Pink rockfish SWFSC ROV TEAM

Swordspine rockfish SWFSC ROV TEAM

Pinkrose rockfish SWFSC ROV TEAM

Rosethorn rockfish SWFSC ROV TEAM

Freckled rockfish SWFSC ROV TEAM

Honeycomb rockfish SWFSC ROV TEAM

Puget Sound rockfish JANNA NICHOLS

Pygmy rockfish SWFSC ROV TEAM

A Comparison of Similarly Appearing Species (continued)

Redbanded rockfish SWFSC ROV TEAM

Flag rockfish SWFSC ROV TEAM

Copper rockfish SWFSC ROV TEAM

Gopher rockfish KEVIN LEE

Black-and-yellow rockfish RICK F

Calico rockfish KEVIN LEE

Brown rockfish DEBBIE KARIMOTO

Grass rockfish CHRIS GROSSMAN

Appendix 2
Counts of Fin Elements, Gill Rakers, and Lateral Line Pores

Note that, traditionally, counts of dorsal and anal spines (as opposed to soft rays) are given in Roman numerals. Counts are given as a range; the numbers in parentheses are the most typical values.

Sources: Hart (1973), Chen (1986), Moreland and Reilly (1991), Poss (1995), Moser (1996), Orr et al. (2000), Nakobo (2002), Love et al. (2002), Orr and Blackburn (2004), and Orr and Hawkins (2008).

ROCKFISHES

Sebastes aleutianus Rougheye rockfish
Dorsal Fin XII–XIV (XIII), 12–14 (13–14); Anal Fin III, 6–8 (7); Pectoral Fins 17–19 (18); Gill Rakers 28–34 (31); Lateral Line Pores 28–34 (31).

Sebastes alutus Pacific ocean perch
Dorsal Fin XIII–XIV (XIII), 13–17 (14–15); Anal Fin III, 6–9 (8); Pectoral Fins 15–19 (18); Gill Rakers 30–39 (33–36); Lateral Line Pores 44–55 (47); Lateral Line Scales 43–55.

Sebastes atrovirens Kelp rockfish
Dorsal Fin XIII, 12–15 (13–14); Anal Fin III, 6–8 (7); Pectoral Fins 16–18 (17); Gill Rakers 28–36 (31–33); Lateral Line Pores 36–50.

Sebastes auriculatus Brown rockfish
Dorsal Fin XIII, 12–15 (13); Anal Fin III, 5–8 (7); Pectoral Fins 15–19 (18); Gill Rakers 21–30 (25–28); Lateral Line Pores 40–50; Lateral Line Scales 45–52.

Sebastes aurora Aurora rockfish
Dorsal Fin XIII, 12–14 (13); Anal Fin III, 5–7 (6); Pectoral Fins 16–19 (17–18); Gill Rakers 24–30 (25–28); Lateral Line Pores 27–31; Lateral Line Scales 41–50.

Sebastes babcocki Redbanded rockfish
Dorsal Fin XIII, 12–16 (13–14); Anal Fin III, 6–9 (7); Pectoral Fins 17–20 (19); Gill Rakers 25–33 (30–31); Lateral Line Pores 41–51.

Sebastes baramenuke Brickred rockfish
Dorsal Fin XIII, 13–14 (14); Anal Fin III, 7–8; Pectoral Fins 18–19; Lateral Line Pores 29–31.

Sebastes borealis Shortraker rockfish
Dorsal Fin XIII–XIV (XIII), 12–15 (13); Anal Fin III, 6–8 (8); Pectoral Fins 17–20 (19); Gill Rakers 27–31; Lateral Line Pores 28–32; Lateral Line Scales 36–46.

Sebastes brevispinis Silvergray rockfish
Dorsal Fin XIII, 13–17 (14); Anal Fin III, 7–8; Pectoral Fins 16–18 (18); Gill Rakers 33–36; Lateral Line Pores 43–53; Lateral Line Scales 58–70.

Sebastes carnatus Gopher rockfish
Dorsal Fin XIII, 12–14 (13); Anal Fin III, 5–7 (6); Pectoral Fins 16–18 (17); Gill Rakers 23–32 (28–31); Lateral Line Pores 34–49.

Sebastes caurinus Copper rockfish
Dorsal Fin XIII, 11–14 (12–13); Anal Fin III, 5–7 (6); Pectoral Fins 16–18 (17–18); Gill Rakers 26–33 (28–30); Lateral Line Pores 35–47; Lateral Line Scales 39–45.

Sebastes chlorostictus Greenspotted rockfish
Dorsal Fin XIII, 11–15 (12–13); Anal Fin III, 5–7 (6); Pectoral Fins 16–18 (17); Gill Rakers 29–36 (31–34); Lateral Line Pores 34–43.

Sebastes chrysomelas Black–and–Yellow rockfish
Dorsal Fin XIII, 12–14 (13); Anal Fin III, 5–7 (6); Pectoral Fins 17–18 (17); Gill Rakers 24–31 (26–30); Lateral Line Pores 35–46.

Sebastes ciliatus Dark rockfish
Dorsal Fin XII–IV (XIII), 13–17 (15); Anal Fin III, 7–9 (8); Pectoral Fins 16–19 (18); Gill Rakers 32–37 (34); Lateral Line Pores 39–54 (45); Lateral Line Scales 44–60 (51).

Sebastes constellatus Starry rockfish
Dorsal Fin XIII–XIV (XIII), 12–14 (12–14); Anal Fin III, 5–7 (6); Pectoral Fins 16–18 (17); Gill Rakers 24–30 (27–29); Lateral Line Pores 37–47.

Sebastes crameri Darkblotched rockfish
Dorsal Fin XII–XIII (XIII), 12–15 (13–14); Anal Fin III, 5–7 (7); Pectoral Fins 18–20 (19); Gill Rakers 28–35 (30–33); Lateral Line Pores 40–51; Lateral Line Scales 48–62.

Sebastes dalli Calico rockfish
Dorsal Fin XII-XIV (XIII), 12–14 (13); Anal Fin III, 6–7 (6); Pectoral Fins 16–17 (17); Gill Rakers 30–37 (34–35); Lateral Line Pores 37–51.

Sebastes diploproa Splitnose rockfish
Dorsal Fin XIII, 11–14 (12–13); Anal Fin III, 5–8 (6–7); Pectoral Fins 17–19 (17–18); Gill Rakers 32–38 (33–36); Lateral Line Pore 32–43; Lateral Line Scales 53–57.

Sebastes elongatus Greenstriped rockfish
Dorsal Fin XIII, 12–14 (13); Anal Fin III, 5–7 (6); Pectoral Fins 16–18 (16–17); Gill Rakers 28–34 (30–33); Lateral Line Pores 37–47; Lateral Line Scales 42–55.

Sebastes emphaeus Puget Sound rockfish
Dorsal Fin XIII, 13–15 (14); Anal Fin III, 6–8 (7); Pectoral Fins 16–18 (17); Gill Rakers 37–43 (39–41); Lateral Line Pores 39–47; Lateral Line Scales 41–46.

Sebastes ensifer Swordspine rockfish
Dorsal Fin XIII, 12–14 (13); Anal Fin III, 5–7 (6); Pectoral Fins 16–18 (17); Gill Rakers 34–40 (36–38); Lateral Line Pores 34–44.

Sebastes entomelas Widow rockfish
Dorsal Fin XII–XIII (XIII), 14–16 (15–16); Anal Fin III, 7–10 (8); Pectoral Fins 17–19 (18); Gill Rakers 33–39 (34–37); Lateral Line Pores 50–60; Lateral Line Scales 58–66.

Sebastes eos Pink rockfish
Dorsal Fin XIII, 11–13 (12–13); Anal Fin III, 5–7 (6); Pectoral Fins 16–18 (18); Gill Rakers 26–33 (27–30); Lateral Line Pores 34–42.

Sebastes flavidus Yellowtail rockfish
Dorsal Fin XII-XIII (XIII), 13–16 (14–15); Anal Fin III, 7–9 (7–8); Pectoral Fins 17–19 (18); Gill Rakers 31–39 (35–38); Lateral Line Pores 46–60; Lateral Line Scales 55–60.

Sebastes gilli Bronzespotted rockfish
Dorsal Fin XIII, 13–15 (14–15); Anal Fin III, 6–8 (7); Pectoral Fins 18–20 (18–19); Gill Rakers 26–30 (29–30); Lateral Line Pores 40–46.

Sebastes glaucus Gray rockfish
Dorsal Fin XIV, 14–17; Anal Fin III, 7–9; Pectoral Fins 18–20; Gill Rakers 34–41; Lateral Line Pores 37–54; Lateral Line Scales 48–78.

Sebastes goodei Chillipepper
Dorsal Fin XIII, 13–16 (13–14); Anal Fin III, 7–10 (8); Pectoral Fins 16–19 (17); Gill Rakers 31–39 (34–37); Lateral Line Pores 48–58; Lateral Line Scales 60–77.

Sebastes helvomaculatus Rosethorn rockfish
Dorsal Fin XII–XIV (XIII), 12–14 (12–13); Anal Fin III, 5–7 (6); Pectoral Fins 15–18 (16); Gill Rakers 27–33 (28–31); Lateral Line Pores 34–45; Lateral Line Scales 42–48.

Sebastes hopkinsi Squarespot rockfish
Dorsal Fin XII–XIV (XIII), 13–17 (14–16); Anal Fin III, 6–8 (7); Pectoral Fins 16–18 (17); Gill Rakers 35–43 (38–40); Lateral Line Pores 49–58; Lateral Line Scales 57–67.

Sebastes jordani Shortbelly rockfish
Dorsal Fin XIII, 13–16 (14–16); Anal Fin III, 8–11 (8–10); Pectoral Fins 18–22 (20–21); Gill Rakers 40–49 (43–46); Lateral Line Pores 52–64; Lateral Line Scales 65.

Sebastes lentiginosus Freckled rockfish
Dorsal Fin XIII, 12–13 (12); Anal Fin III, 6–7 (6); Pectoral Fins 16–18 (17); Gill Rakers 34–39 (36–38); Lateral Line Pores 33–41.

Sebastes levis Cowcod
Dorsal Fin XIII–XIV (XIII), 12–14 (12–13); Anal Fin III, 6–8 (6–7); Pectoral Fins 17–19 (17–18); Gill Rakers 29–33 (30–32); Lateral Line Pores 45–53.

Sebastes macdonaldi Mexican rockfish
Dorsal Fin XII–XIV (XIII), 12–15 (13); Anal Fin III, 7–8 (7); Pectoral Fins 18–20 (19); Gill Rakers 35–42 (37–39); Lateral Line Pores 52–58.

Sebastes maliger Quillback rockfish
Dorsal Fin XIII, 12–14 (13–14); Anal Fin III, 6–8 (7); Pectoral Fins 16–18 (17); Gill Rakers 28–34 (30–33); Lateral Line Pores 34–48; Lateral Line Scales 39–45.

Sebastes melanops Black rockfish
Dorsal Fin XIII-XIV (XIII), 13–16 (14–15); Anal Fin III, 7–10 (8–9); Pectoral Fins 18–20 (18–19); Gill Rakers 32–40 (35–38); Lateral Line Pores 45–55; Lateral Line Scales 50–55.

Sebastes melanosema Semaphore rockfish
Dorsal Fin XIII, 11–12 (12); Anal Fin III, 6; Pectoral Fins 17–19 (18); Gill Rakers 34–37; Lateral Line Pores 34–43.

Sebastes melanostictus Blackspotted rockfish
Dorsal Fin XII–XIV (XIII), 13–15 (14); Anal Fin III, 6–8 (7); Pectoral Fins 17–19 (18); Gill Rakers 30–36 (33); Lateral Line Pores 30–36 (32).

Sebastes melanostomus Blackgill rockfish
Dorsal Fin XIII, 12–15 (13); Anal Fin III, 6–8 (7); Pectoral Fins 17–20 (19); Gill Rakers 27–35 (31–33); Lateral Line Pores 28–34; Lateral Line Scales 32–34.

Sebastes miniatus Type 1 and Type 2 Vermilion rockfishes
Dorsal Fin XIII, 12–15; Anal Fin III, 6–8; Pectoral Fins 17–19; Gill Rakers 33–42; Lateral Line Pores 38–48; Lateral Line Scales 45–48 (Orr et al. 2000).

Sebastes moseri Whitespeckled rockfish
Dorsal Fins XIII, 15; Anal Fin III, 9; Pectoral Fins 17; Gill Rakers 49.

Sebastes mystinus "Blue rockfish"
Dorsal XIII, 15–17; Anal III, 8–10; Pectoral 16–18; gill rakers 32–39; Lateral Line pores 47–56.

Sebastes nebulosus China rockfish
Dorsal Fin XIII, 12–14 (13–14); Anal Fin III, 6–8 (7); Pectoral Fins 17–19 (18); Gill Rakers 25–31 (27–30); Lateral Line Pores 37–48; Lateral Line Scales 43–48.

Sebastes nigrocinctus Tiger rockfish
Dorsal Fin XII-XIV (XIII), 12–15 (14–15); Anal Fin III, 6–7 (7); Pectoral Fins 18–20 (19); Gill Rakers 27–32 (28–31); Lateral Line Pores 36–50; Lateral Line Scales 44–53.

Sebastes notius Guadalupe rockfish
Dorsal Fin XIII, 12–13; Anal Fin III, 6; Pectoral Fins 18; Gill Rakers 35–38; Lateral Line Pores 33–40.

Sebastes ovalis Speckled rockfish
Dorsal Fin XIII, 13–17 (14–15); Anal Fin III, 7–9 (8–9); Pectoral Fins 17–19 (18); Gill Rakers 29–37 (31–33); Lateral Line Pores 45–55.

Sebastes paucispinis Bocaccio
Dorsal Fin XIII–XV (XIII), 13–16 (14); Anal Fin III, 8–10 (9); Pectoral Fins 14–16 (15); Gill Rakers 26–32 (28–30); Lateral Line Pores 51–70; Lateral Line Scales 72–90.

Sebastes phillipsi Chameleon rockfish
Dorsal Fin XIII–XIV (XIII), 12–13 (12); Anal Fin III, 5–6 (6); Pectoral Fins 18–19 (18); Gill Rakers 36–40 (37–39); Lateral Line Pores 29–33.

Sebastes pinniger Canary rockfish
Dorsal Fin XIII, 13–16 (14–15); Anal Fin III, 7–8 (7); Pectoral Fins 16–19 (17); Gill Rakers 38–46 (41–44); Lateral Line Pores 38–47; Lateral Line Scales 43–50.

Sebastes polyspinis Northern rockfish
Dorsal Fin XIV, 13–16 (15); Anal Fin III, 7–9; Pectoral Fins 17–19 (18); Gill Rakers 35–39; Lateral Line Pores 43–53; Lateral Line Scales 69–70.

Sebastes proriger Redstripe rockfish
Dorsal Fin XIII, 13–16 (14–15); Anal Fin III, 6–7 (7); Pectoral Fins 16–18 (17); Gill Rakers 36–43 (37–39); Lateral Line Pores 44–55, Lateral Line Scales 55–60.

Sebastes rastrelliger Grass rockfish
Dorsal Fin XIII, 12–14 (13); Anal Fin III, 6; Pectoral Fins 18–20 (19); Gill Rakers 17–26 (23–25); Lateral Line Pores 40–49.

Sebastes reedi Yellowmouth rockfish
Dorsal Fin XIII, 13–15 (14); Anal Fin III, 6–8 (7); Pectoral Fins 18–20 (19); Gill Rakers 30–37 (34); Lateral Line Pores 47–57; Lateral Line Scales 57–67.

Sebastes rosaceus Rosy rockfish
Dorsal Fin XIII–XIV (XIII), 11–14 (12–13); Anal Fin III, 5–7 (6); Pectoral Fins 16–18 (17); Gill Rakers 27–34 (30–33); Lateral Line Pores 36–46.

Sebastes rosenblatti Greenblotched rockfish
Dorsal Fin XIII–XIV (XIII), 11–13 (12); Anal Fin III, 5–6 (6); Pectoral Fins 16–18 (17); Gill Rakers 28–34 (30–32); Lateral Line Pores 34–42.

Sebastes ruberrimus Yelloweye rockfish
Dorsal Fin XIII, 13–16 (14–15); Anal Fin III, 5–8 (7); Pectoral Fins 16–20 (18–19); Gill Rakers 24–31 (26–29); Lateral Line Pores 39–46; Lateral Line Scales 45–50.

Sebastes rubrivinctus Flag rockfish
Dorsal Fin XIII, 12–15 (13–14); Anal Fin III, 6–8 (7); Pectoral Fins 16–18 (17); Gill Rakers 26–32 (28–29); Lateral Line Pores 39–49.

Sebastes rufinanus Dwarf–red rockfish
Dorsal Fin XIII, 14; Anal Fin III, 8; Pectoral Fins 17; Gill Rakers 37–38; Lateral Line Pores 30–33.

Sebastes rufus Bank rockfish
Dorsal Fin XIII, 13–16 (14–16); Anal Fin III, 5–9 (7); Pectoral Fins 17–19 (18–19); Gill Rakers 31–37 (32–35); Lateral Line Pores 49–56; Lateral Line Scales (89–90).

Sebastes saxicola Stripetail rockfish
Dorsal Fin XIII, 11–14 (12); Anal Fin III, 5–8 (6–7); Pectoral Fins 15–19 (16–18); Gill Rakers 29–40 (31–39); Lateral Line Pores 35–43; Lateral Line Scales 43–53.

Sebastes semicinctus Halfbanded rockfish
Dorsal Fin XIII, 12–14 (13–14); Anal Fin III, 6–8 (7); Pectoral Fins 16–18 (17); Gill Rakers 36–42 (38–40); Lateral Line Pores 40–50.

Sebastes serranoides Olive rockfish
Dorsal Fin XII–XIV (XIII), 15–17 (15–16); Anal Fin III, 8–10 (9); Pectoral Fins 17–19 (17–18); Gill Rakers 29–36 (32–35); Lateral Line Pores 50–58.

Sebastes serriceps Treefish
Dorsal Fin XIII, 13–15 (14); Anal Fin III, 5–7 (6); Pectoral Fins 17–19 (18); Gill Rakers 27–30 (27–29); Lateral Line Pores 44–54.

Sebastes simulator Pinkrose rockfish
Dorsal Fin XIII, 12–14 (13); Anal Fin III, 5–6 (6); Pectoral Fins 16–18 (17); Gill Rakers 28–33 (30–31); Lateral Line Pores 33–40.

Sebastes umbrosus Honeycomb rockfish
Dorsal Fin XII–XIV (XIII), 11–13 (12); Anal Fin III, 5–7 (6); Pectoral Fins 15–18 (17); Gill Rakers 31–38 (33–36); Lateral Line Pores 33–44.

Sebastes variabilis Dusky rockfish
Dorsal Fin XIII–XIV (XIII), 13–16 (15); Anal Fin III, 7–9 (8); Pectoral Fins 16–19 (18); Gill Rakers 32–37 (35); Lateral Line Pores 42–54 (49); Lateral Line Scales 47–64 (53).

Sebastes variegatus Harlequin rockfish
Dorsal Fin XIII, 13–15 (14–15); Anal Fin III, 6–7 (7); Pectoral Fins 17–19 (18); Gill Rakers 36–41; Lateral Line Pores 42–52; Lateral Line Scales 46–58.

Sebastes wilsoni Pygmy rockfish
Dorsal Fin XIII–XIV (XIII), 13–15 (13–14); Anal Fin III, 5–7 (6); Pectoral Fins 16–18 (17); Gill Rakers 36–43 (38–42); Lateral Line Pores 37–47; Lateral Line Scales 45–50.

Sebastes zacentrus Sharpchin rockfish
Dorsal Fin XIII, 13–16 (13–14); Anal Fin III, 6–8 (7); Pectoral Fins 16–19 (17); Gill Rakers 31–41 (34–37); Lateral Line Pores 38–51; Lateral Line Scales 43–59.

THORNYHEADS

Sebastolobus alascanus Shortspine thornyhead
Dorsal Fin XIV–XVIII (XVI), 8–11 (9); Anal Fin III, 3–6 (5); Pectoral Fins 20–23 (21); Gill Rakers 18–23; Lateral Line Pores 29–33; Lateral Line Scales 35–46.

Sebastolobus altivelis Longspine thornyhead
Dorsal Fin XV–XVII (XV), 8–10 (9); Anal Fin III, 4–6 (5); Pectoral Fins 22–24 (23); Gill Rakers 21–26; Lateral Line Pores 28–32; Lateral Line Scales 32–38.

Sebastolobus macrochir Broadfin thornyhead
Dorsal Fin XIV–XVI, 8–10; Anal Fin III, 5; Pectoral Fins 21–23; Gill Rakers 18–21; Lateral Line Pores 30–34; Lateral Line Scales 35–38.

SCORPIONFISHES

Adelosebastes latens Emperor rockfish
Dorsal Fin XII–XIII, 12–13; Anal Fin III, 5; Pectoral Fins 18–23; Gill Rakers 23–27; Lateral Line Pores 28–29.

Pontinus furcirhinus Red scorpionfish
Dorsal Fin XII, 8–9 (8); Anal Fin III, 5–6; Pectoral Fins 17–19; Gill Rakers 15–20.

Pontinus sierra Speckled scorpionfish
Dorsal Fin XII, 8–9 (9); Anal Fin III, 5–6 (5); Pectoral Fins 16–19 (18); Gill Rakers 8–13 (10); Lateral Line Scales 23–29.

Pontinus vaughani Spotback scorpionfish
Dorsal Fin XII, 9–10; Anal Fin III, 4; Pectoral Fins 19–20; Lateral Line Scales 26–31.

Pontinus sp. A. Rosy scorpionfish
Dorsal Fin XII, 9–10; Anal Fin III, 5; Pectoral Fins 16–18; Gill Rakers 13.

Scorpaena afuerae Peruvian scorpionfish
Dorsal Fin XII, 9–10; Anal Fin III, 5; Pectoral Fins 19–21; Gill Rakers 15–16; Lateral Line Scales 24.

Scorpaena guttata California scorpionfish
Dorsal Fin XII, 8–10 (9); Anal Fin III, 5–6 (5); Pectoral Fins 17–19 (18); Gill Rakers 16–19.

Scorpaena histrio Player scorpionfish
Dorsal Fin XII, 10; Anal Fin III, 5; Pectoral Fins 19–20; Lateral Line Scales 24–26.

Scorpaena mystes Stone scorpionfish
Dorsal Fin XII, 9–10; Anal Fin III, 5–6; Pectoral Fins 18–21.

Scorpaena russula Reddish scorpionfish
Dorsal Fin XII, 9; Anal Fin III, 5; Pectoral Fins 20–22; Lateral Line Scales 24–25.

Scorpaena sonorae Sonora scorpionfish
Dorsal Fin XII–XIII (XII), 8–9; Anal Fin III, 5; Pectoral Fins 19–21; Lateral Line Scales 43–49.

Scorpaenodes xyris Rainbow scorpionfish
Dorsal Fin XIII, 9–10 (9); Anal Fin III, 4–6 (5); Pectoral Fins 16–19 (17); Gill Rakers 16–19 (17).

References

Chen, L. 1986. Meristic variation in *Sebastes* (Scorpaenidae), with an analysis of character association and bilateral pattern and their significance in species separation. United States Department of Commerce, NOAA Technical Report NMFS 45.

Eschmeyer, W. N., E. S. Herald, and H. Hammann. 1983. A field guide to Pacific Coast fishes of North America. Houghton Mifflin, Boston.

Hart, J. L. 1973. Pacific fishes of Canada. Bulletin of the Fisheries Research Board of Canada 180.

Humann, P. and N. DeLoach. 2004. Reef fish identification: Baja to Panama. New World Publications, Jacksonville, Florida.

Humann, P. and N. DeLoach. 2008. Coastal fish identification: California to Alaska. 2nd Edition. New World Publications, Jacksonville, Florida.

Lamb, A. and P. Edgell. 2010. Coastal fishes of the Pacific Northwest. Harbour Publishing, Madeira Park, British Columbia.

Love, M. S. 2011. Certainly more than you want to know about the fishes of the Pacific Coast. Really Big Press, Santa Barbara.

Love, M. S., M. Yoklavich, and L. Thorsteinson. 2002. The rockfishes of the northeast Pacific. University of California Press, Berkeley, California.

Moreland, S. L. and C. A. Reilly. 1991. Key to the juvenile rockfishes of central California, p. 59–180. *In* T. E. Laidig and P. B. Adams (eds.). Methods used to identify pelagic juvenile rockfish (genus *Sebastes*) occurring along the coast of central California. NOAA Technical Memorandum NOAA-TM-NMFS-SWFSC-166.

Moser, H. G. 1996. The early stages of fishes in the California Current region. California Cooperative Ocean Fisheries Investigations Atlas No. 33.

Nakabo, T. 2002. Fishes of Japan with pictorial keys to the species. English Edition. Tokai University Press, Japan.

O'Connell, V. 2002. *Sebastes ruberrimus*, p. 248–250. *In* M. Love, M. Yoklavich, and L. Thorsteinson. Rockfishes of the Northeast Pacific. University of California Press, Berkeley, California.

Orr, J. W. and J. E. Blackburn. 2004. The dusky rockfishes (Teleostei: Scorpaeniformes) of the North Pacific Ocean: resurrection of *Sebastes variabilis* (Pallas, 1814) and a redescription of *Sebastes ciliatus* (Tilesius, 1813). Fishery Bulletin 102:328–348.

Orr, J. W. and S. Hawkins. 2008. Species of the rougheye rockfish complex: resurrection of *Sebastes melanostictus* (Matsubara, 1934) and a redescription of *Sebastes aleutianus* (Jordan and Evermann, 1898) (Teleostei: Scorpaeniformes). Fishery Bulletin 106:111–134.

Orr, J. W., M. A. Brown, and D. C. Baker. 2000. Guide to the rockfishes (Scorpaenidae) of the genera *Sebastes, Sebastolobus*, and *Adelosebastes* of the northeast Pacific Ocean. United States Department of Commerce, NOAA Technical Memorandum, NMFS-AFSC-117.

Poss, S. G. 1995. Scorpaenidae, p. 1544–1564. *In* Fischer, W., F. Krupp, W. Schneider, C. Sommer, K. E. Carpenter, and V. H. Niem (eds.). Guia FAO para la identificación de especies para los fines de la pesca. Pacifico Centro-Oriental. Volumen III. Vertebrados – Parte 2. Roma, FAO.

Robertson, D. R. and G. R. Allen. 2008. Shorefishes of the tropical Eastern Pacific online information system. http://biogeodb.stri.si.edu/bioinformatics/sftep/index.php

INDEX

A

Adelosebastes latens	148
Aurora rockfish	50

B

Bank rockfish	12
Black–and–Yellow rockfish	120
Black rockfish	20
Blackgill rockfish	54
Blackspotted rockfish	64
Blue rockfish	26
Bocaccio	6
Brickred rockfish	134
Broadfin thornyhead	144
Bronzespotted rockfish	74
Brown rockfish	126

C

Calico rockfish	130
California scorpionfish	158
Canary rockfish	66
Chameleon rockfish	52
Chilipepper	4
China rockfish	118
Copper rockfish	124
Cowcod	72

D

Dark rockfish	22
Darkblotched rockfish	58
Dusky rockfish	24
Dwarf–red rockfish	100

E

Emperor rockfish	148

F

Flag rockfish	110
Freckled rockfish	90

G

Gopher rockfish	122
Grass rockfish	128
Gray rockfish	136
Greenblotched rockfish	76
Greenspotted rockfish	78
Greenstriped rockfish	106
Guadalupe rockfish	94

H

Halfbanded rockfish	46
Harlequin rockfish	42
Honeycomb rockfish	92

K

Kelp rockfish	132

L

Longspine thornyhead	142

M

Mexican rockfish	10

N

Northern rockfish	38

O

Olive rockfish	28

P

Pacific ocean perch	32
Peruvian scorpionfish	156
Pink rockfish	80
Pinkrose rockfish	84
Player scorpionfish	160
Pontinus	
furcirhinus	150
sierra	152
vaughani	154
Puget Sound rockfish	102
Pygmy rockfish	104

Q

Quillback rockfish	116

INDEX

R

Rainbow scorpionfish	168
Red scorpionfish	150
Redbanded rockfish	108
Reddish scorpionfish	164
Redstripe rockfish	34
Rosethorn rockfish	86
Rosy rockfish	88
Rougheye rockfish	62

S

Scorpaena
- *afuerae* — 156
- *guttata* — 158
- *histrio* — 160
- *mystes* — 162
- *russula* — 164
- *sonorae* — 166

Scorpaenodes xyris — 168

Sebastes
- *aleutianus* — 62
- *alutus* — 32
- *atrovirens* — 132
- *auriculatus* — 126
- *aurora* — 50
- *babcocki* — 108
- *baramenuke* — 134
- *borealis* — 60
- *brevispinis* — 8
- *carnatus* — 122
- *caurinus* — 124
- *chlorostictus* — 78
- *chrysomelas* — 120
- *ciliatus* — 22
- *constellatus* — 96
- *crameri* — 58
- *dallii* — 130
- *diploproa* — 48
- *elongatus* — 106
- *emphaeus* — 102
- *ensifer* — 82
- *entomelas* — 18
- *eos* — 80
- *flavidus* — 30
- *gilli* — 74
- *glaucus* — 136
- *goodei* — 4
- *helvomaculatus* — 86
- *hopkinsi* — 16
- *jordani* — 2
- *lentiginosus* — 90
- *levis* — 72
- *macdonaldi* — 10
- *maliger* — 116
- *melanops* — 20
- *melanosema* — 40
- *melanostictus* — 64
- *melanostomus* — 54
- *miniatus* — 68
- *moseri* — 98
- *mystinus* — 26
- *nebulosus* — 118
- *nigrocinctus* — 112
- *notius* — 94
- *ovalis* — 14
- *paucispinis* — 6
- *phillipsi* — 52
- *pinniger* — 66
- *polyspinis* — 38
- *proriger* — 34
- *rastrelliger* — 128
- *reedi* — 56
- *rosaceus* — 88
- *rosenblatti* — 76
- *ruberrimus* — 70
- *rubrivinctus* — 110
- *rufinanus* — 100
- *rufus* — 12
- *saxicola* — 44
- *semicinctus* — 46
- *serranoides* — 28
- *serriceps* — 114
- *simulator* — 84
- *umbrosus* — 92
- *variabilis* — 24
- *variegatus* — 42
- *wilsoni* — 104
- *zacentrus* — 36

INDEX

Sebastolobus
    alascanus — 140
    altivelis — 142
    macrochir — **144**
Semaphore rockfish — 40
Sharpchin rockfish — 36
Shortbelly rockfish — 2
Shortraker rockfish — 60
Shortspine thornyhead — 140
Silvergray rockfish — 8
Sonora scorpionfish — 166
Speckled rockfish — 14
Speckled scorpionfish — 152
Splitnose rockfish — 48
Spotback scorpionfish — 154
Squarespot rockfish — 16
Starry rockfish — 96
Stone scorpionfish — 162
Stripetail rockfish — 44
Swordspine rockfish — 82

T

Tiger rockfish — 112
Treefish — 114

V

Vermilion rockfish — 68

W

Whitespeckled rockfish — 98
Widow rockfish — 18

Y

Yelloweye rockfish — 70
Yellowmouth rockfish — 56
Yellowtail rockfish — 30